THE BODY
LANGUAGE OF HORSES

THE BODY
LANGUAGE OF
HORSES

by
Tom Ainslie
and Bonnie Ledbetter

*Revealing the Nature of Equine Needs, Wishes and
Emotions and How Horses Communicate Them—
For Owners, Breeders, Trainers, Riders and
All Other Horse Lovers—Including Handicappers*

WILLIAM MORROW AND COMPANY, INC.
New York   1980

Library of Congress Cataloging in Publication Data

Ainslie, Tom.
    The body language of horses.

    1. Horses—Behavior.   2. Animal communication.
I. Ledbetter, Bonnie, joint author. II. Title.
SF281.A36   636.1'083   79-26995
ISBN 0-688-03620-1

Printed in the United States of America

45 44 43

BOOK DESIGN BY MICHAEL MAUCERI

To Hazel Loudon, and in loving memory of Don

PREFACE by Tom Ainslie

We arrived at the horse show just as an exquisitely tailored young man rode his graceful bay over a hurdle. When they turned for the long canter to the next jump, Bonnie Ledbetter told me that the horse would refuse.

Within a yard of the obstacle, the horse skidded to a halt, depositing the young man in the tanbark.

"How did you know?" I asked.

"Body language," she said.

The animal's resentment of its stiff-wristed passenger had been evident to the knowing eye. In a few hours I learned those particular signals and others. I began to make fairly dependable forecasts of my own.

We went to the racetrack. Eight three-year-olds paraded in front of the stands before going to the post. None had ever won a race.

"Look at Number Four," urged Bonnie. "He's more than sharp. He's super-sharp!"

The *Daily Racing Form* record of the colt's previous performances showed that it had raced twice and had been at least fifteen lengths behind the leader at every stage of both processions.

7

"That animal looks prosperous because he wastes no energy in competition," I said, citing the record. "He's a feeder, not a runner."

The colt won with total authority at odds slightly below 25 to 1.

In our subsequent afternoons at the track, Bonnie repeated that kind of stunt several times. My education proceeded. I learned to differentiate the body language of the outstandingly sharp and eager horse from that of the merely fit and ready. I learned to recognize degrees of equine annoyance, from mild irritability to total outrage. I became familiar with the language of fright, from apprehensiveness to abject terror. I can now identify the dull horse, the playful horse, the exhausted horse, the horse in pain, the horse in shock, the horse dehydrated or otherwise debilitated, and the horse so heavily drugged that its ears no longer coordinate with its brain.

We visited a large ranch that specializes in the breeding and schooling of Arabian show horses. Bonnie had not been there before. Within a minute or two of our arrival at each paddock fence, the occupants competed with each other for her murmured compliments and the generosity with which she scratched them in the right places. Some became quite insistent, even jealous. Passing ranch hands raised eyebrows at the young woman nose to nose with livestock. I could have told them something. I knew that Bonnie and the Arabians were exchanging credentials and that everything would be explained in due time.

Horses communicate with remarkable clarity in a language of posture, gesture and sound. They express their needs, wishes and emotions to each other and to the rare human being who understands them.

In describing and translating that ancient language, our book calls attention not only to little-known phenomena of equine nature but to some widely unrecognized infirmities of human nature. For example, horse language demonstrates qualities of awareness unmentioned in prevailing theories of equine intelligence and temperament. By muddling our perception of the horse, those inadequate theories hamper our dealings with the animal. No wonder the typical horse is more frustrated than its keeper, and for some substantial reasons.

Although we naturally direct ourselves to persons who work with horses, we have taken care to accommodate readers whose interest, like my own, is entirely recreational. I can testify that anyone who acquires even a rudimentary understanding of equine language will enjoy horses all the more, being able at last to ascertain an animal's mood before mounting or driving or betting on it. To which may be added the heady pleasure of reading equine minds for one's companions at horse shows, rodeos, Derby telecasts, bullfights and similar assemblages.

Having welcomed all comers, I should point out that the primary concern of our book is the horse itself. The animal's well-being depends directly on the competence of its owners and handlers. With that in mind, we supply an unusual outlook and many techniques indispensable to the development of human-equine relationships more satisfactory than members of either species usually experience.

We began with a review of the horse's faculties, especially its senses, mentality and disposition. This sets the stage for the presentation of its language. When equipped with that basic means of recognizing behavioral and other problems communicated by the horse, the reader will be ready to take corrective measures superior to those prescribed by tradition.

We show how to help a troubled horse by approaching it on its own terms. This avoids the pitched battle so often provoked when, in Bonnie Ledbetter's words, a handler "punishes the horse for being a horse." Equine stupidity is less often instrumental in the downfall of a horse than is the handler's unreasonable notion that the animal should offer more understanding than it gets in return.

After discussing the specifics of troubleshooting, we show how knowledge of equine language applies to the schooling of a young horse—beginning not on the foal's first day of life but in prenatal negotiations with its dam. Another chapter considers the rehabilitation of older horses alienated by cruel or otherwise incompetent treatment.

For sportive types who enjoy Thoroughbred, Standardbred or Quarter-horse racing, we explain how awareness of equine body language helps racegoers in their visits to paddocks and walking rings and in their observations of post parades and prerace warm-ups. That chapter is in two parts. The first part is for casual racing fans. The other addresses dedicated handicappers who may want to understand the varying relationships between a horse's demeanor (which indicates its attitude toward racing) and previous experiences published in its past-performance records.

An appendix gives novel advice on the purchase of horses, including race horses.

So much for our menu and the sustenance it promises. Now for some background. This is Bonnie Ledbetter's book. It consists of her special knowledge and her unique point of view, which derive from a life of deep study, practical experience and, more fundamentally, an abiding respect for the integrity, dignity and sensitivity of animal life.

Many of her views are radically unfamiliar. Yet she actually occupies a position on firm ground, midway between established extremes. For one thing, her mind is free of the sentimentality that attributes human faculties to horses. On the other hand, she rejects the ancient orthodoxy that depicts a successful relationship between human and horse as triumph of masterful man over broken beast. To her, that concept is a self-defeating contradiction. To her, the human-equine relationship is successful only when the horse performs to the outer limits of its possibilities—which it will do only if accepted on its own terms and given options forbidden by orthodoxy.

She has been communicating with animals for as long as she can remember. She has befriended and remained on cordial terms for days, weeks or months with numerous wild deer, one wild bison, many squirrels, certain raccoons and countless birds, including a mama eagle. She has coexisted peacefully with mountain lions and bears. Not long ago, in her twenties and pregnant, she harbored a leopard for months in her home. She says that no wild animal has ever harmed her. Her only difficulties have come from domestic animals disoriented after mistreatment by other persons.

Of all animals, horses have always appealed to Bonnie the most. Some might call them her obsession. She calls them her lifetime study. At Stephens College, in Columbia, Missouri, her equestrian skills defined her as a candidate for international competition. A severe back injury ended that. When her family moved from Missouri to the West Coast, she attended the University of California at Berkeley and then decamped to California State Polytechnic College at Pomona, where the horses were. She was graduated with major credits in English literature, geology and animal husbandry, especially the last.

She has owned, bred, broken (a word she deplores), schooled, handled, lived in barns and pastures with and solved the problems of all kinds of horses. She regrets that she has not yet put a Lipizzaner through its high-school figures in Vienna or navigated an Andalusian in the hair-raising rituals of a Spanish bullring.

Bonnie has time. She is in her early thirties. She lives with her husband and two young sons in Phoenix, Arizona, where we did our preparatory work.

I did not know of her until early 1978, when she wrote to ask me a sophisticated question about the occult art of handicapping horse races. It developed that she had turned to that pastime for recreation only a year or two earlier at Turf Paradise, the Phoenix track, and had become quite adept. So adept, in fact, that she had defeated a leading professional handicapper in a competition conducted by the track management.

She confided that her prowess rested only partly on the formalities of paper-and-pencil handicapping. She never picked a horse to win until she had appraised its physical and mental condition by looking at it in the paddock. Hardly anybody can do that. Few persons even try. When she remarked in another letter that the key was equine body language, I telephoned to find out where she had been all my life.

I told her that I had heard of persons with knowledge comparable to hers but that most of them were long dead. I complained that I was dissatisfied with the sparse literature that had been published about horses' minds. I remarked that many professional horsefolk were oblivious to equine psychology, others indifferent, still others less knowing than they preferred to believe. I had met a few of the best and had found them hungry for insight. So was I.

She told me that all this tallied with her own experience in academic settings, as a member of the racing audience and in years of work with Quarter horses, Thoroughbreds, Hunters, Appaloosas, Standardbreds, Morgans, American Saddlebreds, various draft animals, whatever. She agreed that the fracases that punctuate dealings between horses and handlers are almost invariably caused by defective human understanding and contribute inevitably to the deterioration of the horses. She observed that no amount of ability to recognize and treat swollen fetlocks justified inattention to the many kinds of psychic injury that prevent even sound-legged horses from performing as well as they might.

I insisted that she write a book. She laughed. I offered to facilitate matters by attending to the nuts and bolts. She agreed. I went to Phoenix in February 1979, and remained for six weeks. It could have taken much longer than that for me to assemble and comprehend the material in this book, but her ideas were admirably organized, having been on her mind for so long.

Neither Bonnie Ledbetter nor I suppose that she is the Discoverer of the Horse or that she alone on earth understands its viewpoint and language. At the end of this preface, we acknowledge the important help given her by a supremely expert couple during her years of work on their ranch. No doubt other fully competent men and women are working wonders in barns and paddocks throughout North America. Moreover, it is certain that several English, Irish and Continental horsepersons communicate with their own animals in ways not unlike Bonnie's. Claims of such abilities have been made for bygone Bedouins and various Indians of the North American plains. And the Spanish Riding School in Vienna routinely produces Lipizzaners and riders whose communications verge on the telepathic.

Even if there were no such creature as a Lipizzaner or a cutting horse, or a race rider like Willie Shoemaker to demonstrate that humanity often is correct in its treatment of horses, other facts would remain. For example, people have been domesticating these animals for at least 7,000 years. Until replaced by machinery in the present century, horses were mainstays of agriculture, transport and war. They remain indispensable to private recreations and industrialized sports so popular that the annual revenues of the world's breeders, stable owners, sales agencies, training centers, race tracks and horse-show arenas are measured in billions of dollars.

As that review suggests, we are not here to belittle a history of human achievement. Our objectives lie in the present and future. We offer gain—welcome increases in the pleasures and profits that human beings obtain from horses. Happily, those enlarged satisfactions will be realized only by a dramatic reduction in the needless waste associated with prevailing methods of domestication. Those methods waste horses. In the process, they also waste money.

We promise greater success to persons who acquaint themselves with the language of the horse and the generally reasonable attitudes which that language communicates. We shall show that the animal cooperates most willingly, proficiently and enduringly with the handler who understands it. The results are astounding.

ACKNOWLEDGMENTS

Until good fortune brought her to Don and Hazel Loudon, Bonnie Ledbetter had never met anyone who shared her view of horses. They not only agreed with her abstractly but had been treating horses in that special way for many decades. She worked with the Loudons for several happy years at their Pepper Tree Stables on the old Weisel Ranch in Carbon Canyon, Chino, California. Until then, at schools, colleges and places of work, she had been required to yield to the conventional wisdom, applying her theories surreptitiously whenever she was able to apply them at all. The Loudons encouraged her experimentation and growth. They taught her much. Their ranch was her finishing school. She believes that Don would approve of this book. The first copy will go to Hazel.

Others whom we thank for their generous help are Bob and Misty Strode, Peter M. Howe, Robert E. Tryon, Orlo Robertson, Arthur Ledbetter and Bonnie's close friend, Kat.

15

CONTENTS

1. THE NATURE OF HORSES

In attempting to develop a satisfactory relationship with a horse, it is enormously useful to understand the animal's vocabulary of pantomime and sound. But truly effective communications depend on matters more fundamental than language. To respond adequately to the wants and moods expressed by the horse, we must understand the primary facts of equine nature.

What is it like to be a horse? By means of what physical senses and mental faculties does it perceive and react to its environment? What are its needs and preferences? What are its emotions and how are they provoked? For example, when a horse communicates fear, what is it likely to be afraid of?

Centuries of literature supply a confusion of answers. One reads that horses are hysterically timid, yet courageous in war and sport. One reads that they are nearsighted and color-blind, yet they rivet attention on happenings so distant that a human being needs binoculars to discover what the fuss is about.

We are told that the horse is lamentably dim-witted by comparison with the dog, cat or pig, which possess relatively larger brains. We then learn that the horse finds its way home in the dark over many miles of the most difficult and unfamiliar terrain;

understands and responds accurately to human words and tones of voice; remembers individual persons and significant experiences for years; appreciates diversion, being acutely susceptible to the ill effects of boredom; differentiates vulnerable babies from adults of its own and other species; grieves for an absent friend; exults in reunion; and develops conspicuous enthusiasms and antipathies toward individual persons, animals, activities and inanimate objects.

It is agreed that their senses enable horses to recognize the distant approach of friend or foe, distinguish between the two and signal accordingly. That was why American Indians, who understood the nocturnal utterances of their horses, were usually ready to receive friendly visitors but seldom were caught napping by enemies. That particular equine talent, and an apparent ability to anticipate storms and earthquakes, have aroused scholarly speculation about extrasensory perception. If true, that would be a fancy attribute for a supposedly stupid animal. It might even compel an upward revision of the horse's official I.Q.

In sum, either horses or the published appraisals of them are heavily burdened with contradiction. To indicate that the difficulty lies not with the animals but with the kinds of study to which they have been subjected, let us report and analyze a modest experiment undertaken many summer afternoons ago at the Loudon place in Chino, California.

It was one of those magical hours with chores out of the way, horses drowsing, problems dormant, telephone silent and no visitors to disturb the peace. Don, Hazel and Bonnie took their ease in dappled shade and talked horses. Somebody mentioned equine color-blindness, in which none of them believed. Yet the disability had been described in many texts and was generally accepted, like an article of faith.

Don proposed a test. It would be as objective and scientific as possible. Behind the barn, side by side, were two former oil drums of identical dimensions. One was painted blue with a single horizontal white stripe. The other was green, with exactly the same white stripe at the same elevation. Resting on the drums was a saddle-cleaning rack. The barrels and rack had not been moved for twenty years.

The experimenters removed the rack and reversed the positions of the drums, taking great care. Then they returned the rack to its customary place—same position and same angle relative to the back of the barn, but now atop a different arrangement of drums, with the green one where the blue had been, and vice versa.

Preparations complete, Don strolled around the outside of the barn, entered its front door and went to the stall of Brother Pete, a stallion of Quarter-horse registry, palomino coloring and keen awareness. Don led Pete out of the stall, through the front door, around the side and to the back. When they turned the corner and entered the presence of Hazel and Bonnie, the horse glanced briefly at the women, but something else caught his eye. His ears pricked forward and joined his nose and eyes in deep attention to the oil drums and saddle-cleaning rack. His posture was that of intense curiosity, a primary characteristic of horses.

Pete insisted on going directly to the drums and rack, which he sniffed and nuzzled thoroughly until satisfied that nothing untoward was afoot. He then raised his head and turned toward his friends, ready for work or romp.

The drums and rack had been standard but hardly crucial aspects of Pete's environment ever since his first walk behind the barn many years earlier. Why his sudden interest in those objects? Had he actually perceived a switch of green and blue? Is any other conclusion reasonable?

To illustrate various kinds of undependable thought about horses, and perhaps to immunize the reader against such thinking, we now present and discuss alternative explanations of Pete's behavior. We make no attempt to arrange the explanations in order of plausibility, which they lack.

1. *Pete heard the folks moving the oil drums, wondered what was happening and was eager to find out.*

Although the drums were moved without clatter, the horse undoubtedly heard activity behind the barn. He apparently was unexcited about it. He showed no sign of anticipation or even of interest when led from his stall toward the source of the sounds. Indeed, he behaved exactly like a horse on a routine outing until he turned the corner and, in that instant, became a horse perplexed by a change in the environment.

2. *Being color-blind, the horse could not have differentiated between green and blue, but might have perceived two shades of gray in unfamiliar positions.*

Much of the learned literature in this field of study is polluted by anthropomorphism, an intellectual defect which compels the sufferer to see animals exclusively in human terms. The shades-of-gray theory is not a blatant example but will do. If a horse can distinguish between two colors, it is not functionally color-blind, regardless of whether it sees the colors in shades of gray, purple or pink. The fact that some human beings see gray or shades of gray instead of red or green is entirely beside the point. Horses are not people.

3. *Having overheard Hazel, Don and Bonnie scheming to test his eyesight, the horse had resolved to go along with the gag and pretend that he could differentiate one color from another.*

You may smile but please do not laugh. The sentimental notion that "horses are almost human" is a widespread form of the rankest anthropomorphism. It makes life miserable for any horse subject to its influence. It is the outlook that keeps some horses cooped in their stalls every Sunday so that, in the minds of their owners, they can enjoy God's day of rest—as if the animals were not already half-crazed with boredom and frustration from twenty-three hours of confinement on every other day of the week.

To return to the example itself, horses usually understand large numbers of words associated with their work, but none understands "color-blind." Although capable of flawlessly logical reasoning in many situations, horses do not intellectualize in an abstract way as humans do. They cannot even do arithmetic, although more than one has been trained to enrich its handler by behaving as if it could.

4. *The story does not constitute a sample large enough to warrant the belief that horses can tell one color from another, or even that this particular horse was exceptional in that respect. The whole thing may have been a coincidence.*

A statistically valid observation. Tell it to the owner of a horse that kicks up a rumpus whenever approached by a person in a white coat, having behaved that way ever since its painful encounter with a smocked veterinarian. This reminds us that Don Loudon had a pleasantly tractable horse which turned savage only when Don wore a red jacket. Its previous owner, who had mistreated the animal, always worked in a red jacket.

We have been poking fun at anthropomorphic sentimentalists and other individuals who are unable or unwilling to accept horses as horses. We could enlarge the target by adding those animal-laboratory researchers who imagine themselves to be testing equine intelligence when they challenge a horse to find food in a maze. This test is suitable for a mouse, not a horse. It proves only that, in addition to being nonhuman, a horse is not a mouse. Such tests are among the reasons that so many oracles have underrated the intelligence of horses.

The horse, through eons of genetic adaptation, is a grazing animal whose food grows near its feet and nose. In neither the natural nor domesticated state has it any occasion to push buttons or turn labyrinthine corners in quest of a snack. It has no faculty to assist such an unnatural enterprise and requires none. On the other hand, it can find water in supposedly arid places. Perhaps some researcher will note this revelation, put a thirsty horse at the entrance to a maze, a pail of water in the center, and conclude from the animal's performance that horses are as bright as mice. But the experiment will prove nothing about the horse's intelligence other than that nature has equipped it to find water. In other words, it is a horse.

Hoping that we have encouraged one and all to join us in accepting the horse as it is rather than as one might wish it were, we shall now begin to define it in light of its own characteristics.

* * *

The physical and mental qualities under review here are common to all horses. For our present purposes, innate variations that occur from breed to breed, or among members of a single breed, are less relevant than the similarities. The individuality noticeable in any paddock or stable is a result of both environmental and genetic influences, of course. It reminds us that the horse is among the higher orders of animals. Handlers who recognize and accept equine individuality and modify their methods accordingly are invariably more successful than those who do not. We shall get to that.

PHYSICAL SENSES

Like other forms of animal life, the horse is born with senses of vision, hearing, touch, smell and taste. These prompt its brain to perceive, interpret and react to environmental developments. For the reader's convenience and our own, we shall deal separately with each equine sense. But in real life the horse integrates its faculties, responding to each new stimulus with as many of them as possible. Because its senses of hearing, touch and smell are exquisitely acute, it performs feats that tax human credulity and give rise to theories about extrasensory perception, telepathy and even miracles.

Vision

The placement and contours of the horse's eyes equip it with visual powers admirably suited to life in a pasture or other natural surroundings. If left to its own devices, the animal spends at least half the day nibbling grasses in a nose-down position from which, without even moving its head, it commands a secure view of perhaps 320° of horizon and intervening terrain behind it and on both sides. Naturally, it moves its head occasionally to check the rest of the view. It can see the distant approach of a predator. It can see another horse invading its own sacred territory. It can see its food. It sees what it needs to. Whatever the horse cannot see, it hears, feels or smells.

At work under saddle or between the shafts of a wagon, the horse's remarkably panoramic eyesight continues to inform it of external developments. Its reaction to threats or curiosities perceived in that way are often displeasing to human beings. When pressed to run as fast as it can or jump over high obstacles, a tightly reined horse sometimes rebels against holding its head and neck in positions that prevent it from seeing what it wants to see.

The unhappy rider returns to the barn with the familiar complaint that something spooked the horse, that nothing went right thereafter and, by implication, horsebacking would be more fun if horses were different.

The horse's eyes are on opposite sides of the head and not in front like those of cats, humans and many dogs. Each eye transmits a distinctly different picture through the optic nerves to the interpretive cells of the brain. In general, the left eye surveys objects and events at the left front, left side and left rear. The right eye renders corresponding service on its side. Only when looking straight ahead, with its face perpendicular to the ground, can the animal direct both eyes simultaneously at the same point in space. Some breeds with extremely separated eyes cannot even do that.

Thus the horse views a much larger field than a human being would when standing in the same place and facing in the same direction. When running, a horse can see whether a pursuing animal or vehicle is catching up. In the same moment, it looks ahead for obstacles or other discomforts that might lie in its path.

The animal probably perceives all this with little clarity. Lacking the fused, binocular focus which accounts for three-dimensional vision and accurate depth perception in a front-eyed species like our own, it not only sees things flat but probably in poor detail. However, when sound, smell or sight divert a horse from the overall panorama to a compelling part of it, focus sharpens—the mental emphasis having shifted from routine scanning to a specific point of interest.

Unlike most other creatures of higher order, the horse does not achieve visual focus by moving or flexing the inner lens of the eye. The lens is unadjustable. But the equine retina (site of the optic nerve receptors at the rear of the eye) is unusually irregular in contour. For clear vision the animal raises or lowers its head sufficiently to direct incoming light onto whatever sector of the bumpy retina provides the best image.

For that reason the eyes usually remain stationary in their sockets. Major exceptions occur during intense stress, periods of play or during a yawn. Many readers must remember seeing the eyes of a frightened horse roll in their sockets.

Its eyes being at the sides of its head, the horse obviously cannot see anything close to the center of its face. This is why it backs away or abruptly shifts its head, or both, when an inexperienced visitor to the stall delivers an overhand, downward pat to its forehead. The way to pat a horse is to raise the hand slowly from below and touch the animal's jaw or neck or chin or side of the head before venturing elsewhere. And an unerring test of a horse's trust in its handler is whether the human can place a hand directly on the animal's forehead without provoking withdrawal.

Many authorities have proposed that horses are nearsighted as well as color-blind. Experimental findings support both contentions but not persuasively. We have already touched on color-blindness. As to nearsightedness, it seems probable that most experimentation has failed to consider the paramount role of equine mentality, which is organized to dwell on one thing at a time yet is ever alert to new stimuli and ever prepared for immediate shifts of attention. That attribute undoubtedly helps explain the survival of a species whose first defense against attack was apprehensiveness and whose next was rapid flight. Thus a horse that seems during testing to be visually unaware of a distant object is probably not paying attention to it, being engrossed in the closer and much stranger circumstances of the experiment itself.

Be that as it may, a horse's senses of smell, hearing and touch permit discernment over such long distances that they often initiate attention before eyesight can. But as soon as attention is engaged, eyesight comes into play. A horse notices a coyote so far away that its rider does not see the creature with the naked eye and certainly does not hear it. Whether the horse actually sees it can be left to doubt. The fact remains that the coyote is perceived. The equine senses collaborate with each other to remarkable effect.

Hearing

The loudspeakers of a typical high-fidelity sound system produce musical tones ranging from a low of 20 cycles to a high of 20,000. Anything lower or higher is needless luxury, inaudible to most human beings. The auditory faculties of horses are greatly superior. They hear a much broader spectrum. Moreover, their hearing is many times more acute. They detect approaching footsteps, wheels, whistles, voices and storms long before we can. Indeed, they hear at such stupendous distances that some of us credit them with supernatural powers far beyond the auditory.

The entire body of the horse is a sound receiver. It processes airborne sound through the ears and earthborne sound through an amplifying system which goes from the feet to the inner ear and brain by way of bones, nerves and bodily cavities.

This receiver is always at work. A horse dozes, head down, ears drooping to the sides. A quarter of a mile from the paddock fence, a neighbor's mule crosses the road. The horse's ears spring to attention. It raises its head and directs nose, eyes and ears at the moving animal. Two seconds later, satisfied that all is well, the horse resumes its nap.

A horse's ears swivel on their vertical axes, somewhat like rotary antennas. Most breeds can turn the ears considerably farther than 180°, although none can turn the right ear fully left or the left fully right. This swiveling action facilitates reception. The long concave openings, assisted in many cases by the swinging head, turn directly toward an arresting sound, whether at the front, to either side or directly behind the animal. When the sound provokes apprehensiveness, the horse may swing not only its head but its entire body in the direction already taken by the ears. The ears then assume the forward, pricked position of total interest, all senses working in concert.

The horse's hearing and its interpretations of what it hears are so indispensable to its sense of security that any impairment of this sense causes deep distress. When aural passages are congested by an infection such as a cold, the animal suffers more severely from its anxiety than from the infection itself. For similar reasons, horses stabled near busy airports, rail terminals or other sources of overwhelming noise are seldom as composed and therefore seldom as healthy as they might be. It may be argued that they have impressive powers of adaptation which help them to become accustomed to such uproar. But horses retain the need to attune themselves to their environment by keeping track of its sounds and investigating the abnormal ones. When noise obliterates not only normal sound but normal silence, many horses become flustered. Even if they finally stop reacting to the din with overt anguish, they do not sleep as peacefully or eat as heartily as they should.

Just as disturbed hearing affects equine composure, so does disturbed composure affect hearing. As we shall show, the motions of the ears and the positions in which they come to rest are basic elements of the animal's body language. When profoundly angered, frightened or pained, the horse becomes abnormally indifferent to sound. The positions of its ears reveal this with considerable eloquence. Moreover, in combination with other elements of body language, the ears state the nature of the horse's trouble almost as explicitly as speech could.

There can be no doubt that the ability of riderless horses to go straight home in the dark is at least partly a function of their acute hearing. The same is true of their unerring "forecasts" of electrical storms, high winds or earthquakes. They become restless long before any human being is aware of a change in the weather or vibrations underfoot. And a blind horse uses its hearing and memory to navigate safely within the fences of its paddock or even, as often told, at the front end of a delivery wagon.

It is a great temptation to ascribe equine homing ability, weather forecasts and similar phenomena to some extra sense. In his book *Breeding the Racehorse*, Federico Tesio, one of the most successful breeders who ever lived, speculated about a sense that might enable horses to receive electrical or other radiations from inanimate objects at long distances. He believed that the ears were involved but not in the usual way. Until the faculties of horses are tested more scientifically than they have been thus far, the theory of a sixth sense remains moot. But there can be no doubt of the horse's almost incredibly acute hearing.

Touch

Of all the marvels in its physical repertoire, none surpasses the horse's sense of touch. A well-schooled horse actually seems to anticipate an expert rider's wishes. Such an animal needs neither saddle, bridle nor bit. It receives messages through its skin, and like the barebacked American Indian ponies that dazzled generations of United States Cavalry, it performs the most intricate maneuvers without apparent prompting, as if reading the rider's mind.

Despite appearances, this is no miracle. The human sense of touch being less acute than that of the horse, the rider actually exerts hand or knee pressure before realizing it, but not before the horse realizes it. From a human viewpoint, the phenomenon is roughly comparable to the parlor trick in which a weighted string dangling from a person's hand unaccountably begins moving in circles after having swung back and forth. The muscular activity that induces the circular motion is imperceptible to the person holding the string.

An even slightly tense or apprehensive rider is unaware of the disturbing messages transmitted through knees, hands or seat to an increasingly impatient or demoralized horse. And the muscle language of a confident rider helps to relax and embolden a horse.

Within moments after the birth of her foal, the mare begins communicating with it by touch and voice. It quickly learns to find mental comfort and physical support by leaning against her body, and continues to seek similar security for the rest of its life—like the frightened race horse that leans on the escort pony in the post parade.

The sense of touch being so critical to equine peace of mind, the smart handler employs it as a primary factor in human-horse relations. As we shall show, greetings and subsequent negotiations involve much laying on of hands, including a good deal of solicitous scratching.

We have mentioned the astonishing ability of horses to hear distant sounds. When the sounds are received through the feet, the sense of touch is involved. Touch also comes into play during earthquakes. Horses newly arrived in southern California become jangled by the daily earthquakes to which human beings (but not seismographs) are insensitive. The morbid effects of earth tremors on equine appetite and slumber are at least as responsible as other environmental changes for the poor performances of so many Eastern horses in their first races at Santa Anita or Hollywood Park. By the same token, a California race horse, accustomed to disturbances under foot, may display unease during its first days at an Eastern track where the earth stands still.

Smell

Here again, the equine sense is enormously more acute than the human. The scent of a mare in heat is perceptible to a stallion confined in a stable a half-mile away. He demonstrates this first by lifting his head, raising his upper lip and compressing his nostrils in the toothy grimace known among Europeans as *flehmen*. He then tries to kick down the door of his stall. And the sense of smell is prominent in the horse's recognition of individual human beings, other horses and objects of all kinds. To determine whether someone or something is acceptable, the horse not only looks and listens but does a good deal of sniffing and tasting. Having completed the scrutiny, the horse files the information in its memory. The animal is fairly certain to remember whoever or whatever it encountered, even months or years later. Meanwhile, it goes through life leaving olfactory signs and signals of its own. Its urine and feces are statements to other horses, emphasizing matters of sexuality and territoriality.

In forming acquaintances with its own kind, the horse's calling card is its breath. It exhales through its nostrils into those of a fellow horse, who then returns the courtesy. They now know each other. Human beings aware of this ceremony introduce themselves to a horse by blowing gently into one of its nostrils. If the horse is as kindly disposed as expected, it responds purposefully with its own exhalation. Unless mistreated on subsequent occasions, a horse greeted this way is likely to hold the person in exceptionally high regard, as if it appreciated having been addressed in its own language.

When adding a new broodmare to his band, Don Loudon always took care to introduce her to the lead mare. As the two stretched out their heads and touched noses, the boss mare would almost snort. If the newcomer responded in kind, Don knew at once that she would compete for the lead mare's exalted position. More often, the new mare replied with a gentle exhalation and a low nicker. Don could then turn her loose in the herd, where she would find her own place among the lead mare's followers. Negotiations for standing in the local pecking order always include a good deal of sniffing of each other's flanks, probably to seek out the scent of nervous sweat—the beginning of submission.

Horses abhor the smell of death. One that shies from a trailside bush or resists walking in a direction preferred by the rider is not necessarily perverse, unless the instinct of self-preservation is to be regarded as a vice. Its behavior may be a reaction to the scent of a decomposing animal (too faint for the rider to notice). This scent arouses an instinctive fear that whatever predator killed the animal may still be in the vicinity.

At horse shows, racetracks and wherever else a male horse's libido may distract it from work and embarrass its handlers, aromatic ointments are smeared in its nostrils to mask the scent of females in heat. The reader who considers applying Vicks or a similar substance before trail rides had better ask a veterinarian about the long-term effects of such stuff when smeared repeatedly on a horse's nasal tissues.

Interestingly, when equine senses evolved to their present stage, they left the animals unprepared to smell danger in certain kinds of harmful vegetation. Horses poison themselves on oleander or loco weed. They get intoxicated on marijuana. On the other hand, they avoid poppies, many poisonous berries and dangerous mushrooms.

Taste

When Brother Pete noticed that the blue and green oil drums were newly situated, he was incapable of letting things go at that. It was necessary for him to look the drums over at a safe distance before deciding whether to approach more closely. The final inspection involved considerable sniffing, following by nuzzling with the nose and tasting with the tip of the tongue. For the remainder of his days he ignored the drums, knowing very well how they fit into his scheme of things and how unimportant they were to his survival.

Beyond serving as a medium of identification or instrument of self-preservation (which is one and the same), the equine sense of taste has its more festive side. Lovers of grass, hay and grain though they are, horses are wild about sweets. Molasses is ecstasy. Watermelon rind, as we shall tell later on, never fails as a first step in the conciliation of a horse by an expert human being. And, of course, apples, carrots and peaches are always welcome. Most horses love soda pop and beer. Some get fat when turned loose in walnut groves. Bonnie's ferocious thoroughbred, Moose, adored grapes.

PHYSICAL NEEDS

Yon horse is unaware that several hundred generations of its lineal ancestors have lived and worked in the service of human beings. Except when pastured, they have had no experience in anything like the natural state. When humans think about this at all, they tend to dismiss it as an established fact too obvious to dwell on—an aspect of the evolutionary order of things, the survival of the fittest.

As a statement of the ways of the world, this attitude is beyond challenge. But if we are to promote more satisfactory dealings with the horse, we had better understand that its thousands of years of domestic service have left its fundamental needs and instincts unimpaired.

Barns, stalls, feeding and exercise schedules, saddles, bridles, bits, carts, whips, spurs, jumping contests, running or trotting or pacing races and all the other methodology, paraphernalia and activity of domestication are entirely for human convenience and seldom coincide with equine nature. The horse adapts because it has the ability to do so and, in any event, lacks options. But the handler walks a fine line. When human convenience becomes too oppressive to the horse, it suffers in mind and body. It even dies. This is why superior handlers make frequent concessions to a horse's needs, especially by turning it out to pasture where it can crop grass, sleep under open sky and refresh its spirit in its own instinctive fashion.

In the natural state or any semblance thereof, the horse's principal gait is a leisurely walk. It trots or prances only when showing off. It runs only in play or in flight. It never jumps over anything that it can possibly avoid. It carries no weight on its back. It pulls nothing. It eats when it pleases and spends about half its waking hours doing so. Life in a stable is quite different. No other domestic animal has been compelled to make adaptations as drastic as those imposed on the horse.

With that in mind, perhaps we can accept some of the problems that arise when human convenience frustrates equine need.

Food

Competent stable managements attend adequately to the nutritional requirements of their horses. Centuries of experience have taught professionals all that they need to know about proper types and combinations of hay, grain, mineral and vitamin supplements and treats. Innumerable books and pamphlets and the endless promotional efforts of feed manufacturers inform the profession of new developments in the field. Whether the horse is a livery hack, or a pet that works under saddle three times a week, or an athlete in rigorous training, or a stallion at stud, it probably is fed according to accepted prescription. If the horse eats what it gets, it is sufficiently nourished.

Why, then, do stabled horses raise such a tense clamor at feeding time? And why do paddocked or pastured horses actually endanger the lives of inexperienced handlers who have not yet learned how to bring them their buckets of supplementary grain without inciting jealousy and violence? And finally, in situations where the food itself is both abundant and nutritious enough for physiological needs, why are stable managers as plagued with the problems of bad doers (horses with poor appetites) as good doers are with the whereabouts of the next meal?

Few of these problems would occur as frequently or as intensely if horses lived on rich grazing land, where they could feed themselves at their own slow place for ten or eleven hours a day. But the practicalities of domestication prevent that, and no useful purpose is served by harping on it. Large areas of good grazing are rarely available or affordable. The same is true of the legions of extra human help that would be required to fetch grazing horses to their work.

The stable and its box stalls are here to stay. So are feedings at regular hours. So are racks or nets intended to prevent much expensive hay from falling to the floor. And so, to a considerable, extent, is upsetting tension at feeding time, when many stabled horses are frustrated not only by confinement but by ravenous hunger. The celebrated bad doer that gets so much attention from its handlers may be off its feed because of illness, pain or exhaustion. Or it may simply be dispirited by routine, including the routine of feeding.

Understandably reluctant to lose the services of any horse, profit-seeking stables try to beguile the bad doer. They embellish its menu with extra goodies like sweet feed. Or they feed it more frequently. When those tactics fail, they finally turn the horse out to pasture long enough to restore its equanimity and, with that, its appetite.

Some establishments take preventive measures. They ease the conventional routines with more companionship, exercise or play, all of which we shall discuss later. Another stratagem, totally

impractical with athletic horses, is to let them eat when they feel
like it—and fatten themselves. On the other hand, if a race horse
can be pacified by four relatively small feedings a day instead of
the usual three, wise trainers modify routine and avert larger
troubles.

One standard practice might well be abandoned on grounds
that it causes more troubles than it solves. Many stable operators
place the hay rack so high on the wall that the horse can obtain
the stuff only by biting at it in the manner of a giraffe, with
head elevated. This conserves hay. Horses that obtain their hay
from lower racks drop a lot of it on the floor. On the other hand,
a horse never eats with elevated head. Having reached up to
snatch a mouthful, it immediately lowers its head to a natural
position and begins chewing. Unfortunately, while biting at the
hay with head raised, the horse swallows and/or inhales a lot of
dust and air, inviting digestive and respiratory problems more
costly than spilled hay.

Water

Although the need varies with the season and air temperature
and the kind of work done by a horse, the animal always requires
a lot of water. In places that serve it on rigid schedules in buckets,
an occasional horse becomes dehydrated. Modern stable equipment
includes self-watering basins that are activated by pressure of the
thirsty horse's own nose on a lever or pressure plate.

The old saw about leading a horse to water but being unable
to make it drink is seldom applicable to what happens in stables.
From one end of the world to the other, handlers bar overheated
horses from free access to water. If allowed to do so, a horse in
that condition literally drinks itself to death. This apparent flaw
in its instinct of self-preservation is cited frequently as evidence
of equine unintelligence. What reasonable creature would drink
itself to death?

Undoubtedly, the refusal of a horse to drink itself to death
would make it less of a nuisance and would earn it higher marks.
But, as we have been saying, a horse is a horse. It drinks too

much water when overheated by exercise because nothing in its evolution prepared it for situations of that kind. We have already pointed out that horses do not normally run except in play or in flight. Under human direction they exhaust and overheat themselves by running faster and farther than normal. Nature then bids them restore water to their tissues as quickly as possible, but provides no internal governor to limit the intake.

One more observation about water, equine instinct and human behavior. When provided with unlimited salt, horses lick exactly enough of it to replace what their systems lose in the sweat of exertion. Some stables neglect this need. One sees altogether too many horses with the drawn, tucked-up abdomens and sunken flanks of dehydration. One even sees that form of distress at racetracks, especially in warm weather. In some cases the condition may be due not only to an insufficiency of salt or an inadequate intake of water, but to the use of drugs that increase the excretion of urine. These diuretics prevent the rupture of pulmonary blood vessels and the resultant bleeding to which a small minority of horses are subject when overexerted. The bleeding ends athletic exertion by obstructing the animal's air passages. The reader can decide whether it is better to (a) race a horse weakened by dehydration or (b) retire it because it bleeds.

Sleep

When it feels safe and secure, a healthy adult horse lies flat and sleeps deeply every night. If disturbed by illness, injury, overhead illumination, automobile headlights, noisy trucks or radios or the arrival of a new horse in the barn, horses cannot sleep. Uneasy ones never lie down. This characteristic is built into their genes and is understandable: Recumbent horses are helpless against attack.

Many a bad doer has recovered its appetite and lustrous coat after the dismissal of thoughtless stable hands whose raucous nightly partying made it impossible for the animal to sleep. Persons who seek resemblances between horses and humans will enjoy knowing that, like us, these creatures need their rest.

When their duties permit, healthy adult horses doze at intervals throughout the daylight hours, standing quietly with necks relaxed, heads down, ears flopped sideways, eyes half closed, tails occasionally swinging at flies. In especially safe and quiet surroundings, they sometimes nap in a semiprone position from which they can quickly reach their feet. The forelegs are tucked under the body, hind legs touching the belly. Most of the weight is balanced on the abdomen and one hip. The nose usually rests on part of a curled foreleg. In that position, relaxation may deepen to a light sleep.

Foals and yearlings need more sleep than adults. They stretch flat out on the ground for full slumber at intervals during the day, with mothers or other adults standing guard nearby.

One might imagine that the horse's favorite bed would be cushiony grass. But horses only eat grass and do not willingly sleep on it. They much prefer an open area of dry dirt and, if necessary, they paw and pound at grass until it becomes a suitable bare bed. Their housekeeping is logical enough. They use one area for sleep, another for food and yet another for toileting. They do precisely the same when confined to stalls, and suffer considerably if the floors are not frequently and thoroughly cleaned and the sleeping area dry.

Although a barn provides shelter, horses neither need it nor appreciate it. They merely adapt to it. They are most comfortable under open sky, and sleep more securely when their senses of smell and hearing are unhampered by roof and walls. Confinement to a roofless outdoor pen suits them better. When Turf Paradise built pens of that kind to relieve congestion in its barns, trainers with horses at the Arizona track displayed no more enthusiasm than such construction would have generated at any other track. Human beings prefer to work under a roof. Especially in bad weather.

Exercise

Nothing bedevils the Sunday rider so much as having week-long hopes dashed by the invariable misbehavior of the horse. One of the main causes of the problem is the animal's reaction to lack of exercise during the six days of its rider's absence. Other reasons are listed elsewhere in this book.

Regardless of breed or function, horses need a great deal of exercise. They need it daily. They need it for physical tone and mental health. When paddocked or pastured, they exercise themselves. When stabled, they require the assistance of human handlers and do not always get it.

Saddle horses need not exercise under saddle, but they languish without frequent opportunities to wander around and inspect things while stretching their legs. Having languished too long, many of them turn sour and balky.

Even if ridden daily, a young horse that lives in confinement provides a better ride if allowed to run loose for fifteen minutes beforehand. It dashes in circles by itself, bucks exuberantly and finds a nice place to roll. After that is time enough for the grooming and saddling. This routine would probably have small effect on the performance of a horse that is cooped up for days on end. For that, the only solution is either to move the horse to pasture or establish a program of daily exercise, the expense and/or inconvenience of which are repaid in the animal's health and tractability.

All horses love to roll. The ritual is more than exercise. It is essential to their own hygiene, with which we will now deal.

Grooming

Friendly members of an equine herd pair off to groom each other's skin and coat. They nibble away at dead skin and loose or matted hair, relieving bothersome itches, freeing pores for the healthy passage of sweat, and removing foreign matter which might dull the skin's vitally important nerve endings.

Mares groom their foals frequently and vigorously. Besides helping to keep the little ones clean, this deepens their social and emotional reliance on the sense of touch. A bit later, when the youngsters begin to lose the fuzzy fur of infancy (an itchy process), they groom each other and are eager candidates for the friendship of human beings willing to scratch them.

Throughout their lives in stable or pasture, horses continue to depend on grooming for physical comfort and mental repose. When grooming is neglected by stable help, a horse tries to find relief by abrading its skin on walls, posts and fences, often damaging them and itself and, worse, creating difficulties with its handlers, who then accuse it of vices. In more competent hands, conscientious grooming increases rapport with the horse, builds its confidence and is a powerful aid to successful training.

Even when groomed by a handler and at intervals by another horse, the individual animal reserves certain grooming rights for itself. No delight exceeds its joy in rolling, preferably in mud but happily in sand or dust. The friction soothes its skin while rubbing off loose particles, absorbing sweat and fluffing matted hair. Emerging from a mud wallow, the horse looks dreadful to the uninformed human observer, but it is thoroughly relaxed. After the mud has dried to powder, the animal completes the process by rolling in dirt and then shaking off most of the dried mud. The treatment is comparable to a beauty parlor mud pack and is as effective as an hour of thorough grooming by a human using a stiff brush.

The need to roll is most intense after heavy exercise or long confinement. A hot horse that is allowed to roll in sand rids itself of sweat and relaxes more rapidly than one whose handlers prefer to walk it around in a cooler blanket and give it occasional sips of water. If deprived of rolling at such times, some horses become neurotic. Even when dead tired, they sleep fitfully. Even when physiologically hungry, they eat poorly.

A very special treat for some horses is the privilege of rolling in a stream, on the edge of a lake, or in surf. The lucky animal lies on its back in the shallow water, legs in air, wiggling and groaning with delight. Then it clambers upright and shakes off the water like a dog. The urge is so tempting to a hot, tired horse that experienced riders know what to expect when they stop near a stream or shore. Inexperienced ones get impromptu baths.

Horses that crave more rolling than they get are sure to attempt it in their stalls. One price paid by the management of an improperly tended stable is the death of a horse cast in its stall. That is, when trying to roll, the animal jams its folded legs against a wall and is unable to muster the leverage needed to regain its feet. Hampered breathing and impaired circulation lead quickly to paralysis and death.

Two additional notes about rolling. At tracks where authorities routinely test the urine of horses to see whether performance has been affected by illegal drugs, the functionaries who try to obtain the urine sample often wait hours before a tired animal produces any. But a racer that is allowed to roll before cooling out is likely to arrive at the testing barn in a relaxed state. It cooperates promptly.

Finally, some horses can roll on one side at a time but not from one side to the other. To reach the other side, they must climb to their feet and then return to the ground on the desired side. A more athletically limber horse performs the entire ritual without interruption. Shrewd purchasers of horseflesh value that ability.

Space

We have mentioned the fastidious equine need for separate spaces in which to eat, sleep and excrete. Like other needs of an apparently physical character, an unsatisfied need for space becomes a mental problem. Its most frequent manifestation occurs in an ill-managed barn with inadequately cleaned stalls. Horses respond to this kind of deprivation in the same way as to others—with bad spirit and poor performance.

Whereas some horses accommodate readily to confinement in a reasonably maintained box stall, it can be stated with certainty that all horses prefer paddocks to stalls and pastures to paddocks. Extreme cases never really adapt to prolonged confinement, even after thousands of years of it. The signs are especially pronounced in a typical mare near foaling time. Feeling the need for more space than she has, she suddenly chases other mares away from her and even threatens the life and limb of her old chum, the ranch dog. When human convenience dictates that she complete her pregnancy in a barn stall, she pines. She is not a robot. She does much better when allowed out in the weather, where she is sure to find a sufficiently dry spot to lie down and deliver more naturally and more comfortably.

Horses with conspicuously developed senses of territory are hard to handle. Their evil tempers brand them as rogues. Even allowed free run in a pasture, they drive off all approaching horses and humans and charge savagely at birds that commit the offense of landing on their turf. It can be assumed that some horses are born that way. A more plausible assumption is that many become so in response to deprivation.

Later on, we shall see that effective training of a troubled horse requires keen alertness to and respect for its need of space. Poor results await an inconsiderately eager or aggressive person who comes too close too soon, invading private space without an invitation from the horse.

MENTAL AND SOCIAL ATTRIBUTES

Now we come to the center of things. The brain of the horse is more competent than is generally supposed. Its capacities include long memory; sensitive comprehension; alert curiosity; surprising resourcefulness; considerable sociability, such as playfulness; solicitude; selective friendliness; and even a sense of injustice.

In support of these assertions, we offer facts that are readily substantiated wherever horses exist. These facts proclaim that the animals are able to give more than the average rider or handler knows how to get. Their generous response to reasonable treatment is a function of their mentality and disposition. So is the balkiness and sourness and general deterioration with which they respond to unreasonable treatment.

We are suggesting that the differences between human and equine minds are neither as wide nor deep as tradition insists. But, to reiterate a central emphasis of this book, the difference remains extensive. The horse is not, as sentimentalists say, "almost human," and it defies comprehension unless seen as what it is—a horse.

Fears

Of all domestic animals, the horse is most often overcome by hysterical fright. The fright reaction is part of its evolutionary baggage—its sensory perceptions and instinct of self-preservation. Some stimuli arouse fright in virtually all horses. Individual animals develop specific fears of their own. Unfortunate experiences leave a conditioning imprint on the horse's brain. Which is why expert handlers take pains to avert experiences of that kind and to neutralize them when they occur. We shall now discuss fears to which all horses are susceptible.

Loss of Balance: Because it is helpless against attack when off its feet, a healthy horse only lies down if it feels safe. For the same reason it hesitates to risk loss of balance on unfamiliar footing. Confronted with a wooden bridge for the first time, a normal horse becomes apprehensive. It needs to test the strange material with its feet before committing its weight and risking its balance. With a rider intelligent enough to dismount, the animal checks out the surface, allows itself to be led back and forth over the bridge, and then willingly crosses with weight on its back.

Similarly, a horse resists its first experience on a mountain trail strewn with frighteningly loose rocks. If not allowed to accustom itself to the new challenge, it may be unable to complete the trip without injuring itself and the rider. Even if the horse eventually overcomes the fear of rocky trails, that first bad experience will have affected its general attitude.

Fear of falling also accounts for the miserable performances of many race horses in their first competitive efforts on muddy or sloppy tracks. No basis exists for the familiar theory that one never knows how a horse will race on a slippery surface until it has tried. Wise trainers know better. They turn their horses loose on muddy training tracks—giving them ample opportunity to become acquainted with the peculiar footing. The horses then accept the weight of exercise riders for additional schooling, including the experience of breaking from the starting gate on a slippery day. If the animal can run in mud or slop, the trainer now has no doubt about it. By the same reasoning, winning stables train young horses in mud caulks (cleats that help the shoes achieve footholds on slick, wet surfaces) before expecting the animals to race with the things. No count exists of the number of young horses totally ruined by the horrendous experience of running in mud for the first time in their lives not in a relaxed training session but under actual racing conditions—and while wearing mud caulks for the first time. Under those handicaps, the youngster struggles for balance, probably cuts one of its legs with an errant, caulked hoof, and is lucky not to fall. The memory lingers.

When attacked by a mountain lion which drops onto its back from a cliff or a tree, the horse's best defense would be to fall backward and crush the predator. But no horse does so. Instead, it tries to flee. Why, then, does an angry horse deliberately rear and fall backward on an inept rider? The difference is that the horse recognizes the lion's intent to kill and is instinctually incapable of falling to the ground. Instead it tries to knock the lion against trees as it runs, bucks, whirls and rears. But it perceives a human being in another light. The rider is not on its back to kill. So a horse angered by some characteristic of the human being is able to risk a brief fall to get rid of an annoying rider.

The Unfamiliar: Until satisfied that something new is harmless, a horse tends to assume that it may be dangerous. We have seen that its attitude toward a new footing relates to its profound fear of falling. But it displays comparable apprehensiveness in a new environment, or when startled by an unfamiliar sound or smell or appearance of an unfamiliar moving object. Unless the handler suspends operations and helps the animal to deal promptly with the cause of its fear, problems multiply. For example, one conventional way of teaching a horse not to shy at windblown pieces of paper is to crack it across the chest. Rather than be hurt, the animal may stop shying at paper. But it continues to fear that phenomenon and, in the bargain, nurtures resentments against the persons who beat it. A much more rewarding approach is to stop work at once, pick up the offending piece of paper, ball it up and bat it around to show how harmless it is, and then let the animal smell it and notice that the handler's own familiar scent is on it. After a few demonstrations of that kind, the horse stops shying in any bothersome way. Some become so confident that they shy not at all, merely avoiding pieces of paper and going about their business. Later chapters contain numerous other illustrations of the principle that it is always better to work with the horse in terms of its current level of development rather than punish it for its nature.

Predators: A horse may never have seen a snake or a big cat in its life, but is sure to fear either on first sight. In fact, the mere scent of a big cat is enough to cause panic. If circus horses tolerate tigers, it is because during years of training their nostrils have been crammed with aromatic salves which obscure the offending scent. Horses are also skittish about large dogs of predatory aspect. They look uneasily at German Shepherds and Doberman Pinschers but quickly accept Collies. Toy Poodles and other small dogs are almost beneath notice. If one comes within range, the horse might sniff at the wiggling ball of fluff just to see what it is. And a gentle Great Dane elicits great curiosity. The horse plainly regards it as something not quite dog and not quite pony, but usually does not fear it.

The Poll Bone: This is a loose bone between the ears. If struck with force it can penetrate the brain and kill the horse. Instinctively protective of this vulnerable spot, horses may become frantic when led toward a low-roofed van or trailer, particularly if they have hit their heads on one before. In those circumstances, some handlers actually equip the animals with helmets to protect against injury and, hopefully, reduce fear.

As a dubious last resort in attempting to teach manners to a resistant horse, dramatic effects are obtained by striking the animal over the poll bone with a fragile wine bottle filled with a slush of sand and warm water. When the bottle breaks and the warm, moist substance dribbles down its head, the savage horse becomes a trembling wreck. A gestured threat to repeat the treatment is usually enough to terminate subsequent misbehavior. The extreme maneuver with the bottle wins the horse's attention and renders it manageable but scarcely forms the basis for a fully productive relationship.

A horse stands tranquilly in its stall, head outside the gate. A well-meaning tourist approaches and tries to pat it between the eyes. To see the hand, the horse raises its head and, if tall enough, hits the door frame with its poll. The experience makes the animal head shy and unreceptive to future pats when standing in its stall. The same effect is produced by handlers who hit horses on the head to stop them from nipping.

Water: Horses rarely learn to love swimming and seldom enter deep water for the first time without great apprehensiveness. A rider who wants to take a horse across the first stream of its career can achieve that purpose in one of two ways. The first is coercion and leaves its mark. The other takes no longer and is more constructive: The rider dismounts and shows the horse that streams are fun. This involves certain inconveniences, such as getting into the water, splashing around and making festive noises. If the animal has already learned to trust the person, which is probably the case with someone who handles fears in this way, it may now dip its hoof into the water without physical urging. The rest follows easily.

Death: We have mentioned the reluctance of horses to approach a decomposing carcass, the scent of which is more apparent to them than to their riders. The sight of death is just as unwelcome. If a horse dies in a pasture, the other occupants stay as far away from the corpse as they can possibly get. They probably associate all death with predatory attack, fearing that the killer may still be near.

Less easily accounted for is the hysterical flight of a horse into its burning stable instead of away from that peril. Horses often die that way, overpowering their handlers as they rush toward the flames and smoke. On the other hand, they avoid fires that break out elsewhere. Besides identifying a difference between the intelligence of the horse and the human being, this awful phenomenon may be a byproduct of domestication. Stables do not exist in the horse's natural habitat. But for domesticated horses they are home. Maddened by the screams of burning horses and the overwhelming noises made by fire-fighting humans, the rescued horses burst free and rush home.

Remembered Fears: If a horse has not been helped to overcome a fear, it will forever after be rattled, or worse, when confronted by the cause of the fear. The more often the experience is repeated and corrective measures are omitted, the deeper and more difficult the problem becomes. Equine memory is awesome.

Memory

On the third day of Sissy Fashion's life, a handler was working in the pasture. The foal teetered on one side of her grazing dam. The handler was nearby on the other side. Something disturbed the mare. She began jumping around. Sensibly getting out of her way, man and foal collided. The foal fell flat. Six years later, Sissy Fashion was herself a broodmare on the same ranch. In fact, she was the leader of the broodmare band, ruler of the roost. Whenever the man who had knocked her down entered the pasture, Sissy tried to maneuver herself into position to nail him with a hoof. She hated him. She had hated him passionately since the age of three days. Otherwise, she was a placid and friendly mare.

Don Loudon used to insist that his horses were easy for him to handle because they remembered him from the earliest days of their lives as a figure of superior size and strength. He could have added that he was unfailingly gentle with foals and that they accepted him as a trusted friend as well as a superior. Later, when he mounted a horse born and raised on his ranch and became the first rider it had ever carried, he provoked none of the bucking, squealing and associated hullabaloo so familiar to viewers of cowboy movies. Instead, the horse would turn its head in astonishment to see what in the world its friend was doing up there.

Having no bad memories of its association with Don, a young horse could accept his weight without resentment or fear. But a horse raised with less concern for its sensitivities is quite a different creature. Just as an unfortunate child flinches whenever an adult male raises a hand, a horse reacts reflexively to any sight, sound, touch or smell that reminds it of a bad experience. Although the memory may have been imprinted years previously, the merest reminder instantly jangles the horse.

During a recent season at Turf Paradise, a Thoroughbred that should have been winning some of its races was finishing no better than third and frequently much rearward of that. In every race it slowed its stride a few yards before leaving the final turn and entering the home stretch. It even did this during its morning workouts. In accordance with the standard practices of racing, its riders were instructed to whip the animal on the turn "to keep its mind on its business." This did not work. The horse would shorten stride, whipped or not, and would not resume full effort until it found itself on the straightaway that led to the finish line. It was never able to overtake the leaders.

To solve a problem of that kind or any other, it helps to know the specific cause. Failing that, it is essential to know the categories of unpleasant memory into which the cause might fit. Discussing this horse's costly behavior with its baffled trainer, Bonnie suggested that the animal had undergone a harrowing experience on some turn for home at some track where it had raced or exercised during its checkered career. Perhaps it had been knocked into the rail, or had stumbled, or had run into a fallen horse, or had injured itself in a hole.

She assured the trainer that horses are seldom creatures of whim, being engrossed in the basic practicalities of their often difficult lives. She remarked that the rider's whip was not distracting this horse from its fears. Instead, as the consistency of the Thoroughbred's behavior and the disappointing results of its races indicated, the abuse was simply adding physical discomfort to mental. She recommended that the jockeys stop beating the animal and concentrate on encouraging it with voice and hands during the turn for home. Having little more to lose, the trainer decided to try the new tactic. After three weeks and three more losing races, the animal apparently became accustomed to positive experiences on the turn. One day, instead of slowing its pace, it continued running and won.

Thus, unpleasant new experiences generate bad memories and knotty problems, but good new experiences have the opposite effect. A horse forced to enter a trailer, even though frightened by the unfamiliarity of the vehicle, will become terrified of trailers and will resist them until someone takes the prolonged trouble to reeducate him (which may not be possible if the pattern is too deeply ingrained). But a horse properly introduced to this strange new contraption is conditioned to accept it as a normal feature of its life. Should something untoward then happen to the horse while in the trailer, the results will not be catastrophic. The previous conditioning will hold firm. We shall produce details in a later chapter.

It must be evident by now that the horse's keen memory helps it learn the skills of its particular trade—especially when intelligently trained. To appreciate that memory and deal with it in a constructive spirit is easier, more natural and infinitely more successful than to assume that the animal is a perverse rattlebrain which understands little except coercion.

Finally, it should be kept in mind that a horse's memory is operational at all times—at leisure as well as at work. Some of the few moments of accuracy in sentimental motion pictures about horses are those that portray the animal's pleasure when returned to a favorite farm or reunited with a dear friend after months or years of separation.

Curiosity

The horse's curiosity is as keen as that of the cat or raccoon, and its need to satisfy the feeling is no less acute. The main difference, of course, is that the domesticated horse spends most of its time in confinement or other restraint and cannot inspect what it wants to unless it can win permission from its handlers. Moreover, the horse's emphases are mainly defensive. Its first concern is survival.

It therefore needs to be thoroughly familiar with the details of its surroundings. Its curiosity quickens at the sight, sound, smell or touch of each new object, moving or not, animate or not. What is this new activity? What is that in the sky? Anything unfamiliar must be identified and catalogued. If permission to do this is denied, or the urgency of the need to do it is expressed in a manner that draws punishment from an impatient or ignorant handler, trouble brews. Not only does the animal's work suffer during the incident but the human-equine relationship worsens in consequence.

If a horse were incapable of making mental associations among perceived phenomena—categorizing the swooping hawk as harmless—it would spend its life in panic. But it does associate and, for all practical purposes, it categorizes. This is a sign of intelligence as significant in the horse as in its master. That it sometimes takes the horse a minute or two to satisfy curiosity and add to its catalog is a price that the wise master pays. And a privilege that the contented horse repays.

Ridden onto a strange track for its first exercise, a spirited horse stands still and surveys the scene. Belying the rumors of its inadequate eyesight, it looks high and low and in various directions. It sniffs the air. It listens intently. All at once it moves forward under its rider, having made the necessary associations. In the same circumstances, another horse may not delay the proceedings in that way. It may be a dull animal to begin with. More likely, it has learned from painful experience not to interfere with human schedules and, to the degree that it suppresses its curiosity, it is a less spirited and less eager horse. In fact, it is less horse.

Brother Pete noticed the new arrangement of oil drums behind the Loudon barn and satisfied his curiosity in a few seconds. If the drums had been new to the scene and, more alarmingly, the first drums of his career, he would have been more elaborate about his inspection. Having directed his full attention to them, he would have circled the drums slowly, backing occasionally to see if they might launch an attack. When reasonably sure that they lacked aggressiveness, he would have approached with great care and no great haste. At last, he would have craned his neck until the bones almost popped and would have brought his muzzle near the objects for analytic sniffs and licks.

As the reader already knows, equine curiosity is a tax on human patience. And human impatience tempts human improvidence. The rider taking a horse past a new trailer will see and feel its head turning toward that object. The rider wants to be elsewhere, the horse needs to linger. To berate the animal as stupid and, in extreme cases, to punish it for its needs are acts that prevent it from being a horse. A horse treated that way cannot be as good a horse as it might under more enlightened auspices. The rider gets back to the barn on time but arrives there with not quite the horse that set forth earlier in the day.

Playfulness

When they are three or four weeks old, foals are no longer compelled by instinct and maternal direction to spend all their time with their dams. They begin to play with each other like puppies. They romp in circles. They race from one fence to another. They rear and buck and snort. They play king of the mountain on the farm's manure pile, racing up, rolling down, clambering back.

The spirit of fun persists with age and growth. It is a key factor in the successful schooling of a horse. The young horse that is allowed to regard work as fun learns its skills most rapidly and applies them most willingly. When opportunities for play are sparse, the animal devises games of its own. It plays with objects in its stall. And it often invents pastimes with which to bother human beings.

When hectoring a clumsy handler is the only game in town, healthy horses take shrewd advantage. Not long before we began work on this book, Bonnie watched a frustrated trainer trying to slip a bridle onto the head of a perversely playful horse. As soon as the bridle and the man's hands approached, the horse threw its head up, out of reach. After several repetitions the red-faced trainer looked despairingly at Bonnie.

"Have you any molasses?" she asked. Most barns use it to sweeten feed. The man brought some. Bonnie rubbed a handful on the bit and walked to the horse. It raised its head. She reached up so that it could smell the molasses on her hand. The horse lowered its head, took the sweet bit avidly and the bridle was in place.

"I'll be go-to-hell," said the man.

Not all human beings are as patient as that trainer. Some react angrily to the antics of a playful horse. Some would have hit the horse with the bridle, instantly converting a neutral item of equipment into a feared weapon. The physical punishment would have ended the game but would have created new difficulties.

A horse's favorite playmate is another horse, but it relishes play with its handler. They chase each other around a tree or across a paddock, playing a kind of bump tag that sometimes leaves the handler sprawling as the animal struts off with its tail in the air. In a relationship advanced enough for that kind of activity, nobody gets hurt. At a word, the horse slows its pursuit to a considerate trot. No horse should be deprived of play. One whose need for play is wholesomely satisfied will not concoct perverse games of its own.

Jiminy Cricket loved to play with a gunnysack full of tin cans. When a human audience convened at the paddock fence, he would sniff at the sack, paw at it, nicker at it, seize it with his teeth and trot around the ring with it dangling noisily from his mouth. Then he would break into a lope. When he reached the part of the fence diametrically across the paddock from the audience, he would take dead aim and fling the sack at the folks.

As we have stated and shall demonstrate in our chapter on schooling the young horse, game-playing is fundamental to equine education. Not at all surprisingly, the animals use it as a technique in teaching their own young. Whether they think of its educational features in terms as abstract as cause-and-effect is doubtful and irrelevant. Yet consider a horse called Willie.

Willie was an aging gelding of mixed ancestry and cumbersome physique who had become the official baby-sitter at the Loudon place. He spent his afternoons in the yearling pasture, fussing over the youngsters and devising games which amused him and, accidentally or not, helped to prepare the yearlings for later life. Each day at almost exactly 3:50 P.M., when the yearlings had been napping and would soon awaken for play, Willie would lurch among them, pause and lightly nip one on the rump. Its slumber rudely disturbed, the baby would leap squealing to its feet, rousing the others, while Willie trotted around in high amusement. The game toughened the spirits of the yearlings, as each proved by reacting less vociferously when nipped awake a third or fourth time. More important, it taught the growing youngsters the potential dangers of being caught napping in daylight.

The old gelding's capacity for playful pestiness took him late each afternoon into an adjoining pasture occupied by a small herd of cows and their tough bull. To do this, he had to jump uphill over a five-foot fence, no small effort for one of his seniority and size. He always did it when the cows were resting. He would intimidate the bull into a cranky retreat, then herd the cows to their feet and force them to walk all the way around the hill and back. He would then take his leave. One day, the owner of the cows complained to Don Loudon that someone or something was taking weight off his beeves. The problem was solved by moving the herd to a pasture on a distant section of the ranch.

Boredom and Frustration

The two-year-old filly had always been sweet and tractable. She was still cooperative on the day she returned from a race with several deep cuts on her legs. To protect her against further hazard while the cuts were healing, the trainer kept her in the stall for a full week.

When she was well enough for exercise, they took her out and saddled her. The rider mounted and headed for the track. The filly did not want to go. She fought like a wild horse. It took the rider half an hour to complete what was normally a two-minute stroll. When they finally reached the entrance to the track, the filly reared and flipped backward to the ground. The rider was nimble and lucky. He escaped without injury and promptly quit his job.

One week of confinement had altered the filly's disposition. We have already discussed the equine need for regular exercise and have mentioned the equally basic need for diversion, which is not always the same thing.

The filly would have gone willingly to the track for a strengthening gallop if her handlers had understood the psychological effects of a week's confinement. They should not have rushed her back to arduous work before giving her a chance to enjoy her release from the stall. But they treated her conventionally. That is, they overlooked the difference between a horse and an automobile. A repaired car goes where it is driven. But the horse has a brain that generates feelings and preferences.

The psychology of the horse is that of a browser. It does best with daily changes of scene. It needs to be taken from its stall as often as possible and, as they say out West, "messed with." It wants to poke its nose into things, feel different textures beneath its feet and see different sights. If deprived of these diversions, it becomes bored and frustrated, even when given ample exercise under saddle (which most race horses get). To relieve the monotony of twenty-three hours in the stall and one hour of hard work, grooming and cooling out, horses do the best they can. They tend to develop infuriating vices, such as trying to knock down their stall doors with their hooves, or gnawing on them with their teeth, or nipping at humans who pass within reach.

It follows that a horse given adequate diversion is one that is readier for work and more composed in the stall. To accept this reality, one need not overestimate or romanticize the powers of equine mentality. One simply must recognize that the brains of living creatures regulate their behavior.

Comprehension

Horses understand each other very well. When a pastured broodmare approaches food or territory already appropriated by the leader of the broodmare band, the leader commands the trespasser to back off. A mere lowering of the ears delivers that message.

An itching horse obtains relief by going to a chum. After salutations, the needful one identifies its itch by gently nipping the corresponding area of its friend's hide. They then groom each other.

Equine comprehension of human wishes is even more impressive. Although the terms of a horse's employment invariably require it to render services incompatible with its own natural disposition, it frequently understands what is asked and learns how to comply.

Unless the horse is an unusually dull animal or has been turned off by mistreatment, it understands the words that describe the gaits and other physical maneuvers of its work. It understands other words that refer to aspects of its behavior, approved or otherwise. It understands and reacts to tones of voice. It understands the meanings of various pressures on its mouth, neck and sides. It understands the significance of many situations.

If the prevailing situation seems uncongenial or threatening, the horse expresses its reservations with full candor. For a good example, consider the two-year-old filly that fought her rider rather than work out after a week of confinement (see *Boredom and Frustration* above). She associated the saddling and the weight of the rider with a stressful activity of which she wanted no part at the time.

Her behavior indicated the normal limits of equine comprehension. A young horse understands but resists human demands for cooperation in any undertaking that strikes it as a threat to its own survival. As often as not, the horse's fears are groundless. The human is right. The animal's survival is not at stake. But the horse cannot comprehend that. If spirited, it fights back and suffers for its intractability. If less spirited, it surrenders. In either case its relations with its handlers deteriorate and so, in time, does its own quality as a horse.

Which brings us once again to a central truth already emphasized. Because a horse's powers of comprehension are limited by comparison with those of its handler, the handler who wishes to address the animal constructively can do so only within the animal's own frame of reference. Had the balking filly's people been more concerned with her long-term well-being than with winning an argument, they might have understood what was on her mind. They would then have taken her for a leisurely walk and let her browse and roll. They would have spared her the psychological and physical rigors of a workout until two or three days of diversion and light exercise had helped her recover from the distress of confinement.

Why did they not accept that easier and more productive alternative? Why did they prefer to traumatize the animal in a fight? Perhaps they were unaware of their options. More probably, they did not care to lose dignity in the ancient battle for supremacy over a supposedly pea-brained species. So they pushed a perfectly decent horse around the bend. At this writing, the filly has not yet raced as well as she did before her confinement. What price dignity?

Another element of equine comprehension resembles our own sense of injustice. One manifestation involves the famished hunger typical of horses fed at regular intervals in their stalls. Here again we encounter the instinct of self-preservation. If the handler deviates from routine, bringing the first bucket of grain to a horse that is usually fed fifth, the others create a huge fuss. Or if a handler has established a welcome routine by passing a carrot or some other special treat to a horse for five successive days but omits the delicacy on the sixth, the horse may sulk. Indeed, if no carrot materializes on the seventh day, the horse may refuse to look at the handler on the eighth.

Similarly, a human being arouses jealousies among paddocked horses by offering more friendly attention to one than to others. If the others enjoyed the person's attention in the past and have come to expect it, they become jealous. They may crowd the new favorite and, if still frustrated, may try to nip the human visitor.

The horse's ability to put two and two together is loudly demonstrated in any barn after a handler takes out one horse for a change of surroundings and is slow to return for the others. The neighing, stamping of hooves and kicking of walls qualify as emphatic messages. Experiences like those inspire certain horses to become escape artists, unlatching their own stall doors and rushing out.

Don Loudon once had a veritable Houdini of a stallion which learned how to slip the stall bolt at will. To prevent this, Don fastened the bolt with heavy baling wire. A week later, Bonnie heard the door open and saw the stud rushing off the premises with a piece of the wire dangling from his mouth. The other studs in the barn set up a frantic kicking and bellowing. They did not enjoy being left behind.

Don's next tactic was more complicated. He not only wired the bolt but fastened the end of the wire to a nail in the wall, several feet beyond the horse's reach. The next time the horse escaped, the broken wire was found hanging from the nail.

Reasoning that the horse needed diversion, Bonnie wondered if he might be appeased with an activity other than lock-picking. Perhaps he just wanted something to fumble with. She got a soft rope and tied knots at short intervals along its full length. She gave it to the horse, who pawed and sniffed and mouthed it for a while and then began untying it with his teeth, one knot at a time, five minutes per knot. When all the knots were gone, the stud slipped the door bolt, walked out and presented the rope to Bonnie. He then returned to the stall. This time she hung the knotted plaything from a nail at the rear of the stall. Thereafter, when he wanted more knots, the stallion would neigh for someone to come and retie the rope. He tried no more escapes from the barn.

Speaking of the equine ability to put two and two together, many horses learn to understand words other than "Trot" and

"Whoa," or deeds other than the sting of a whip. An example occurs in stables that discourage undue rumpus by paneling the upper halves of the stalls with a heavy wire mesh that carries a light electrical charge. When handlers are at work in the barn, they switch off the electricity. If a horse becomes restless and begins kicking things, someone usually can restore decorum by walking toward the electrical panel and saying, "All right, I'm going to turn it on."

Another sign of equine comprehension is the ability to discern differences between human adults and babies or small children. An occasional public riding stable is lucky enough to own an old warrior that specializes in transporting kids. Let a small child shift its weight sideways in the saddle and the horse shifts its own to restore the youngster's balance and prevent a fall.

Bonnie recalls that some horses found great pleasure in playing with her son Bradley, when he was still an infant in arms and she took him to a stable. Attracted by baby sounds, several horses extended their heads over the stall doors for closer looks. One nickered with particular interest. When Bonnie went to its stall and presented the baby, the horse sniffed at him, tickling his belly with its nose. The baby giggled in delight. The horse rolled its head joyfully and tickled the laughing baby again and again.

Well, now. We have only scratched the surface of equine intelligence, but we have probably built a sufficient foundation for the material yet to come. It must be clear that horses not only know more than many persons realize, but are capable of learning more than they already know. From the heaviest, most phlegmatic draft horses to the most delicate miniature breeds, one finds surprising levels of comprehension. As in any other species, some individuals are brighter than average and others are subnormal. But horses as such are receptive and responsive animals.

They learn from each other. They learn from humans. They learn by example. They learn from experience. They retain what they learn. When thoughtfully managed, they learn to be the happy, confident partners of trusted human beings.

Social Attributes

Contrary to all motion pictures about noble stallions, mares are the dominant members of the wild herd. As dams, teachers and protectors of the foals, their influences are the first and most enduring in the life of every horse. Colts never forget the authoritativeness of the adult female, including the pain she dispenses when pestered. When free to choose her own mate, a mare easily intimidates an unwanted stallion and runs him off.

A fight between stallions features a great deal of noisy posturing and ends when one retreats before suffering serious damage. A fight between stallion and mare ends the same way—with the stallion in retreat. Similarly, a stallion may drive a predator away from the herd, but a mare will pursue it with intent to kill.

Besides transplanting horses from their natural habitat to stables and paddocks, domestication abolishes the herd. Except briefly at breeding time, mares and stallions are segregated from each other. Thus females become not only the first but the sole adult influence on foals.

Domesticated mares seldom remain in charge of their offspring for as many months as they would in the natural state. Commercial breeding establishments wean the youngsters on schedules calculated to spare the energies of broodmares already pregnant with their next foals. In larger plants, male weanlings go into one pasture and fillies into another. The twain meet only in passing, unless selected for breeding purposes as adults.

After countless generations in environments so different from those of their origin, domesticated horses are by no means the same animals that their ancestors were or that their distant wild cousins still may be. Selective breeding for various forms of sport and work has modified their physiques and, to some extent, their minds. But millions of years of genetic endowment are not easily obliterated. The horse remains a highly gregarious animal whose inherited need for company is reinforced by experiences at the side of its dam during the crucial first months of life.

Pastured, paddocked or stabled, horses seek each other's friendship and solicitude and offer as much in return. Horses in adjoining stalls manage physical contact with each other. When one is led out for exercise, they call to each other. In the welcome spaces of a pasture, they form closer associations based on play, mutual grooming and the solidarity of companionship.

Although horses tend to associate in pairs, they demonstrate a larger loyalty to the group as a whole and, in a real sense, to the species as a whole. A newly arrived horse—especially a stallion—undergoes a period of probation, which may include physical hazing, before finding its own place in the local hierarchy. On the other hand, after gaining even tentative acceptance, the newcomer benefits from what can only be described as instruction. The classic example is that of the electrified fence which inflicts disciplinary shock on any horse that tries to break through or burrow under it. By their own actions, the senior residents show that the fence should be avoided. The newcomers follow suit with unerring consistency.

Another phenomenon: In the rigorously structured hierarchy of the broodmare pasture, exceptions are made for a blind mare. When she calls her foal, other mares chase it back to her. They surely have no way of understanding a concept like blindness. But they understand each other's communications. The blind mare, unable to search for her missing foal, calls for it with particular insistency while remaining in one place. Her voice and body express great tension and anxiety, which are not lost on the other residents of the pasture. If the foal does not return to the dam by itself, other mares restore peace and quiet by herding it back to her, often using impatient body language to hasten the youngster on its way.

In larger bands a placid old-timer sometimes volunteers to adopt a newly orphaned foal. She invariably is low on the local pecking order and leads a comparatively sedentary life. She almost surely is uncomfortably full of milk. The bawling of the orphan upsets and disorients the other babies, annoying the lead mare and other higher-ranking members of the band. But the noise strikes a responsive chord in our old-timer, whose own foal is not taking as much milk as she can give. She nickers to the motherless one. It rushes to her, slams its body against hers and devours her dripping milk. This relieves her painfully full udder. Order is restored to the pasture.

The lead mare wins dominance by physical and psychological means. She rules as long as she remains vigorous. Her powers serve twin purposes—first choice of food and space (a) for herself and (b) for her young. By natural selection, the other mares organize in declining order of priority, with the lowest and most subservient getting the last and least for herself and her foal. Unless the pasture is inhumanely crowded, everyone subsists. But the psychological effects on the foals are substantially important. As Number One in its own age group, the lead mare's baby becomes habituated to the deference of its peers and their dams. If well bred, soundly constructed and not too severely disoriented by premature weaning, the Number One foal emerges as Number One weanling, most likely to succeed in what humanity calls the Game of Life.

Although horses naturally prefer each other's company, they welcome human friendship at every stage of their lives from the day of birth. Moreover, when cared for by knowing persons, they flourish even if totally deprived of equine companionship. In such circumstances or any other, a normal horse responds to human affection with simple trust, loyalty and keen sensitivity. The movie that depicts a horse's resourceful concern for the well-being of its injured master is, for a change, quite accurate. For that matter, so is the one in which a distressed horse rejects the attentions of bad men but willingly accepts help from a sympathetic stranger. Like Androcles's lion, a horse can usually tell when someone really wants to help.

2. HORSE LANGUAGE

A short vocabulary of body language gives full expression to the horse's limited range of needs, wishes and emotions. Any interested person can learn the basic patterns quickly and easily.

Having absorbed the material in this chapter, readers will discover that a certain amount of practice is needed before the lessons can be applied with full success. The main difficulty is one of changing the ways in which we habitually receive communications. Having spent our lives trying to understand other human beings, we are accustomed to listening to words while surveying the facial expressions that sometimes speak louder than words. This will not work with a horse. Its bony face is not nearly as expressive as ours. In time, we learn to read things in its eyes, but by then we also have learned that its communications are fully understandable only when our gaze encompasses its entire body.

From ears, eyes and mouth to legs, feet and tail is a vast expanse. If we tarry too long at the ears, we may miss the whole recital. Similarly, if we yield to human instinct and dwell too deeply on the horse's vocal utterances, we are lost. The snorts, nickers, neighs, whinnies, bellows, squeals and screams of horses are meaningful only in conjunction with the body language of the moment or the situation of the moment, or both. The sounds are best understood as punctuation marks or emphases.

The Happy Horse

In a moment of particular enjoyment, the horse drops its head and then flips it high and describes a full, skyward circle with its nose. Between times, its behavior is eager, interested, alert, playful and responsive.

The well-known horse laugh sometimes contains elements of derision. It involves the same rolling gesture of the head, plus a curling of the upper lip, full exposure of the upper front teeth and a good deal of prancing with raised tail. When old Willie frightened the sleepy yearlings, his pleasure expressed itself in that way. And Jiminy Cricket did it after scattering his human audience with a flying gunnysack full of tin cans.

A raised tail is the primary sign of playfulness. The tails of playful youngsters curl right over their backs. The gesture initiates a game. One yearling hoists tail. The other replies in kind. Then they raise their heads, arch their necks with muzzles down, prick their ears forward and take off at a dead run. When especially exuberant, the youngsters launch themselves by rearing, spinning on their hind legs and racing away.

A race or show horse that approaches competition strutting with a raised tail is happy about the whole thing, including the noise of the crowd.

The Proud Horse

When thoroughly pleased with itself and the prevailing situation, the horse registers its satisfaction by prancing with ears straight forward, nostrils flaring, tail up, head pointed downward on arched neck. At the canter, it raises its hooves with easy co-ordination and stretches its legs in a smooth, eager stride. The neck remains arched, the head down. This is the body language of a horse whose pride is well founded. Mares display it when introducing their new foals to the herd. A stud cordially received by mares compliments himself with the same extravagantly gorgeous strut.

The guileless candor of the horse is most charmingly demonstrated in its proudly pompous self-satisfaction. Horses seldom try to mislead other horses, and they do not intentionally mislead human beings. But they sometimes fool themselves. The strutting stallion discovers after a close encounter that another stallion is even noisier and more pugnacious. It runs off and stops strutting until the psychological wounds heal. Or the same stallion is cut down to size by an unreceptive mare.

An even more interesting case of self-deception is the horse that struts like the monarch of existence without impressing many other horses. One of the reasons that it impresses few horses is that its body language differs from that of authentic pride. Instead, it has about it the air of what Bonnie calls "a cool cat." Its eyes are narrow and knowing, its prancing smoothly economical. It looks about unexcitedly, almost disdainfully, as if matters were well in hand. The proud horse, far more outgoing in its jubilant body language, performs more successfully in competition of any kind. The cool cat usually comes unstuck under stress.

The Interested Horse

Whether interested in some familiar proceedings or overcome by curiosity about something new, this horse conveys total absorption. Nose, eyes and ears point straight ahead at the object of interest, and the animal moves its body directly into line, if allowed. We have already described the elaborate backing and filling and circling of curiosity.

The Eager Horse

The eager jumper loves the game and is anxious to get on with it. He minds neither the crowd noise nor the sight of the high fences. He shows his impatience by stamping a front foot. And then a hind foot. He shakes his head. He dances sideways.

His behavior often is confused with that of fractiousness or nervousness. But he is not actually unruly and is nervous only in the keyed-up way of good performers before they take the stage. His fundamental composure is apparent to anyone who notices that he sweats relatively little, whereas an apprehensively nervous horse is more likely to drip lather. And the fractious horse remains fractious when called upon to go to work. But the eager horse heads enthusiastically for the show ring or starting gate. As soon as the waiting is over, he settles down beautifully and approaches the job with full relish. He takes the first jump a foot higher and several feet longer than necessary in exuberant relief that the game has begun.

And his attention to the work is total. His ears turn completely back to his rider, or forward to where he is going.

The Healthy Horse

The glowing coat reflects a lot of light. Mane and tail are soft rather than matted, lumpy or coarse. In summer the hindquarters are dappled with the intensified color of health. The animal eats eagerly and finishes all the feed. It tends to be playful, responsive to attention and interested in everything. Finally, its hooves are strong. They grow vigorously and rapidly, and are less susceptible to cracks than those of a less prosperous horse.

The Sharp Horse

Whether its work is some form of athletic competition or simply outdoor recreation for its owner, a horse is described in this book as "sharp" when its body and spirit are at peak condition. Its body language combines eagerness with the glowing appearance of health. Unless competing with more experienced or intrinsically superior animals, or unless derailed by a frightening experience or a clumsy handler, the sharp horse can be expected to perform more satisfactorily than it did before sharpness set in. As our description implies, a horse may be healthy without being especially eager, or vice versa. But when the two combine, look out.

The Bereaved Horse

When her foal is taken away before she is ready to wean it, a mare suffers grief and worry. The fretful behavior and body language are unmistakable. Any horse displays these signs after sudden separation from a favorite companion, such as another horse, a stable dog or a beloved owner.

The animal spends some of its time moving around its stall in restless circles, and much more with its head out the door, ears pricked forward, looking in whatever direction the longed-for companion was last seen. The horse constantly sniffs the air for the missing scent. It frequently utters long, nickering calls. If the separation is more than brief, the horse goes off its feed, loses sleep and does its work poorly.

In the same plight, a pastured horse spends its time racing up and down beside the fences, looking and calling. Other horses may return the call, but none of theirs is the wanted voice.

The Frightened Horse

Apprehensiveness: The horse directs full attention at the object of concern. Ears, nose and eyes point straight at it, although the head may swing slightly or the animal may step sideways for sensory focus at different angles. A good deal of loud sniffing is probable, even if the object is at a distance. Furrows appear over the eyes, giving them an appearance of anxiety that is completely recognizable when combined with the rest of the body language. Occasionally, the horse withdraws its head, with neck highly arched. But the eyes, ears and nose remain on target. The horse becomes light-footed, dancing restlessly in place.

Acute Nervousness: When an apprehensive horse is unable

to satisfy its concern and discover that it actually is safe, the signs
intensify. If the problem is unspecific, as when a horse finds itself
in noisy new surroundings, the movements of head and ears become uncoordinated. The head turns one way, the ears flick
from one direction to another, trying to capture a significant
sound. Unable to face the threat as it normally would—because
it has not been able to isolate and identify one threat as more
menacing than the general situation—the animal dances in circles.
As its undifferentiated fears and frustrations increase, so does
its urge to circle. If forced to face in only one direction, its feet
continue to move. Its tail swishes aimlessly, now on the vertical,
now from side to side. It begins to shiver. If it did not sweat during its period of apprehensiveness, it does so now, especially over
the kidney area between its hind legs, possibly on the flanks and
neck as well.

Thorough Fright: A horse does not willingly tolerate prolonged fear. To rid itself of the sense of menace, it tries to flee.
Its instinctive wish is to rear on the hind legs, pivot and rush
away. Its tail becomes still, and the battle with the handler begins.
The human being pulls in one direction, the animal raises its
head to retreat in the opposite. The handler pulls more vigorously. The horse plants its feet and at the slightest opportunity
spins away.

A merely nervous horse avoids fences, but a truly frightened
one goes right through them and may step on humans or slam
into other horses. If it breaks free, it runs with nose extended,
ears turned backward to the source of fear—bound for the barn or
wherever safety lies. But if restrained from flight, it may become
angry and seek to throw its rider or kick its handler. Otherwise it
loses itself in terror.

The horse is now completely awash in sweat. The whites of its
rolling, frightened eyes are probably visible. If led away by a
mounted handler, it moves diagonally in a high-stepping, thoroughly uncoordinated gait, its head over the other horse's withers
for comfort, or perhaps held unnaturally high by the escort rider's
lead chain—a position that gives the handler greater leverage and

the horse less. The abject animal breathes heavily, with much nostril movement. And it emits frightened nickers in a voice so low that it seems to be talking to itself.

Total Panic: When nothing is done to allay its terror—(such as removing it from the place of fear), and when the situation is sufficiently prolonged, a susceptible horse goes beyond fear to the border of insanity. Its eyes glaze. It screams. If it now breaks loose, it may try to go through a barn wall. Or it may tear itself to pieces on a barbed-wire fence. Or it may collapse in shock, falling flat on the ground, where it lies shaking from head to toe and breathing heavily.

Panic expended, the horse is in a state of nervous exhaustion. It stands with feet widely spraddled in an effort to maintain vertical balance, head almost to the ground, eyes closed, ears flopped down motionless to the sides, body quivering, scarcely breathing. It takes no food or water. It may remain in that condition for minutes or hours. Two days are fatal. If the handlers try to use a rope or other means to force the animal upright or into a walk, it usually staggers and often falls flat.

The Bored Horse

When tired of waiting for the next activity or when denied sufficient exercise or diversion, the horse displays its boredom with eloquent signals. It shifts its weight restlessly from foot to foot. It holds its head sleepily downward for a while and then moves it actively, with much ear motion—hunting up something of interest. For lack of anything else to do when among other horses and their riders, it may turn around and chew on a rein or on a string hanging from another horse's saddle pad. Or it may nip a horse to arouse it or perhaps to alarm the human beings and get something started. Bored horses are extremely mouthy.

In its stall a bored horse fiddles restlessly with whatever it can find, which usually is not much. It may contrive to break the valve lever of its self-watering appliance or open the latch of the stall door. In time if its needs are unfulfilled, it usually develops a vice, becoming a cribber, a weaver, a wind-sucker or a noisy lip-flapper—a general pain in the human neck.

The Sour Horse

"Used to be I could get him going with barely a touch of the spurs, but now I can really crack him and nothing much happens. I think he's sick."

Or: "I can't get him to do anything. He doesn't eat much either. Doesn't act angry. Just doesn't do anything. The vet says he's not sick."

Those were frustrated owners describing horses whose spirits had soured. Such horses resent barns, stalls, cars, people, exercise. They want nothing further to do with their routines or with the persons who supervise the routines. Their fuses are burning short. In time their sour undependability becomes total unco-operativeness. In more time the animals become dangerous.

A sure indication of incipient sourness is expressed in unmis-takable snubs. As soon as the horse sees its handler entering the barn, it turns to face the rear of the stall. If permitted, it remains in that position, presenting its rear end to the unwanted human being for the duration of the visit. At this early stage the same horse may treat other human beings with interest, not yet having rejected all members of the race as equally undesirable.

The most common sign of deepening sourness is grudging ex-penditure of effort. A jerk on the bit, and the ears fall back slightly or the tail slaps the rump—signs of impatience. Grudging performance soon becomes conspicuously substandard. The ani-mal responds less quickly, less energetically, less powerfully; above all, less generously. And then comes fractious balkiness, the hallmark of a sour reluctance to cooperate.

A dedicatedly sour horse does not really cooperate even when whipped. From the former slight tensing of the ears and the quick, irritable half-slap of the tail, it strikes a posture in which its ears lie backward, approaching the pinned-flat position of out-right anger. Its body is tense. The muscles of the jaw bunch in knots, as if the teeth were clenched. The upper lip quivers. The eye has a hot and angry look.

When a handler arrives to clean its stall, a barn-sour horse flattens its ears and lifts a rear leg slightly off the floor as if to kick. Or it swings its head menacingly, preparing to bite. It no longer is eager for food. Whoever brings the food sees the ears flatten, the muzzle curl, the tail slap. The horse does not touch the food until alone once more. Indeed, it follows the handler's departure with its eyes, to make sure the person is not coming back.

Of all sour horses, the most dangerous are mares. One that nurses a long-standing grudge may seem peaceful enough to the world at large, but the detested handler had better not blink when within kicking range. And if the object of resentment happens to be another horse, male or female, the sour mare will make a career of intimidating it.

The Angry Horse

Having described sourness as a chronic state of resentful balkiness in which overtly aggressive anger is readily provoked, we will now consider the body language that expresses various degrees of anger. A horse need not be sour to become furious. A perfectly tractable animal may be enraged by unfair or coercive treatment or by the nagging of some other horse.

For example, a nice young colt had been allowed to pass the time in play with the operating lever of his automatic watering plate. He finally broke it. His handler now tied it down to prevent him from flipping it. When the colt managed to break the tie, the handler whipped him. The mood instantly switched from fascination with the metal device to anger at the handler. The same kind of thing may happen if a change of stable routine forces a horse into lengthier confinement and less outdoor play than it has been accustomed to. Or trouble may arise if a horse is forced without advance preparation to practice walking through unpleasant-feeling rubber tires in rehearsals for a Western pleasure show. The more abrupt and less understandable the change, and the more coercion involved, the greater the anger may be. And if the anger provokes unjust punishment, the handler can expect a real fight.

Irritation: The ears lie back lightly. The tail swishes to one side, as if at a fly. The hindquarters tense, ready to kick.

Greater Irritation: The ears lie closer to the skull. The tail motion is more vigorous on one side. The rear hoof on that side now raises slightly off the ground.

Thorough Irritation: The ears now are pinned down. The tail lashes from side to side. The eyes flash anger. The horse kicks straight back at the source of irritation.

Outright Anger: The tail whistles horizontally, moving quickly enough to cut human skin and draw blood. Occasionally, the tail pops straight up and down, slapping against the rear. The head turns toward the object of anger, ears pinned, upper lip curled or quivering (a sign of intent to bite), eyes aglare and taking dead aim. Having zeroed in, the horse shifts its weight to the front and kicks with both hind legs. If this does not suffice, it may spin on a rear hoof and leap forward, its head lowering and then stretching toward the target like that of a striking snake, mouth wide open. The teeth crack shut with the force of a bear trap. Or the attack may be an attempt to strike with the front hooves. Throughout all this, the horse does not sweat a drop. Neither does it utter a sound. When the voice finally comes into play, it emits the drawn-out squeal heard on movie sound tracks during staged battles between stallions.

Mad Fury: When the situation persists and anger is too long unresolved, the horse crosses the threshold of insanity. Its angry squeal becomes a deep bellow. Its eyes seem glazed and unfocused. It begins to sweat heavily from its physical exertion. It flings itself bodily at the source of anger, or at anything that may obstruct its way to the target. Its sense of self-preservation suspended, the animal very probably hurts itself and may even break its neck and die. If not taken in hand and pacified, it finally collapses in shock and exhaustion resembling the state to which severe fright so often leads.

The Horse in Pain

Mild Discomfort: When bothered by a fly sting, the chafing of a saddle or cinch (girth), the beginning of an arthritic condition in a joint, or a pebble in a hoof, the horse responds with a variety of countermeasures. It shrugs or shivers an affected muscle, hoping to dislodge the irritation. When that fails, it stamps the nearest foot. Or it slaps its tail as close as possible to the discomfort. If still unrelieved, it swings its head backward to hit the offended part with the side of its muzzle. Then it makes for the nearest wall, tree or post and tries to scratch itself. If a leg is involved, the horse tries to rub it or nibble at it. Unless prevented, it can chew an arthritic knee to shreds.

Bothered by discomfort short of real pain, the horse is unusually active. When its head is not engaged in biting or bumping at the sensitive spot, it bobs up and down in constant disquiet. The motion is especially persistent when discomfort occurs in the head itself. To rid itself of a disagreeable bit, it tosses its head repeatedly. If the problem is at the side or back of the head, the animal shakes its head and ears in a lateral or rotary motion, often in both.

Afflicted in these ways, a horse is unable to concentrate on other things. The effects are particularly noticeable when the site of discomfort is out of reach, like an intestinal gripe or an abrasion beneath the saddle. If expected to carry a rider, the horse almost surely complies, yet shows no willingness to go places. It wants to stop and concentrate on its problem. Asked to gallop, it shortens stride.

Greater Discomfort: The first time a horse feels the whip, it behaves as if stung by a bee. Its ears turn backward to the site of the sensation, its tail slaps at the spot and it jumps aside, away from the bad feeling. No horse takes long to learn the connection between whip and rider, an insight that scarcely enlarges the animal's affection for that individual. Although whips are widely and frequently used by riders of all degrees, and some equestrians feel absolutely disrobed without them, the implements are seldom employed intelligently. In most hands they do more harm than good. For example, the blow of the whip invariably produces sideward motion, but does not frequently achieve the immediate forward accleration that is sought. For better effect and less chance of collision with fences or nearby horses, a good rider simply shows the whip, bringing it along the animal's side from rear to front in an overtaking motion from which the horse is happy to run away.

Back to bodily discomfort itself. When the sensation is internal, as in the beginning of an attack of colic, the animal exudes melancholy. Its ears turn back toward the discomfort. Its eyes look blank, as if switched off. Engrossed in its problem, the horse scarcely flinches when a hand passes in front of its face. Its rear legs shift frequently, its abdomen moves up and down. It turns occasionally to touch its side with its nose. And it groans.

Severe Discomfort: Most of the conditions just discussed are capable of worsening, whereupon the horse's body language becomes more pronounced, more insistent. Soreness is one discomfort that is likely to become severe before attracting much equine or human attention. When a sore foot or leg really bothers a horse, it prefers to stand still, trying to keep weight off the affected member. It holds the foot an almost imperceptible distance from the floor, but off the floor it is. You can slide a piece of paper under it. The muscles of the unaffected leg bunch under the extra load. An experienced observer can differentiate between this posture and the toe-down, hip-shifting body language of drowsiness.

Pain: When led forward, the sore horse must put weight on

the hurting foot or leg. Moving as slowly as possible, it burdens the affected leg as lightly and briefly as it can. As the foot approaches ground, the head bobs downward on the other side to maintain a balance that enables the animal to spare itself extra discomfort. As pain mounts, the horse carries its head lower than usual, and the limp becomes more pronounced. In severe or persistent and unrelieved pain, the limp becomes a three-legged hop. The injured limb never touches down. The ears flop to the sides, lying parallel to the ground like airplane wings. The eyes are glazed and unresponsive to a hand that passes in front of them.

When required to move out at a trot or canter, a horse in moderate pain displays absolutely no eagerness. It expends as little energy as possible, favoring the hurt with a slow, choppy, awkwardly uneven stride, ears sideways in the airplane position, head down and bobbing. If the soreness is behind rather than up front, the ears are swiveled toward the pain and rarely turn in any other direction. Sore in any foot or leg, the animal finds the trot most painful, the canter less so. Interestingly, if the pain is simple soreness rather than the tissue damage of a ruptured tendon sheath or bucked shins, the heightened flow of blood during exercise may actually "warm" the horse out of soreness, allowing it to perform comfortably. This likelihood increases if the animal is treated with considerate patience and given much verbal encouragement during that uncomfortable process.

A horse afflicted with the severe pain of colic demonstrates the problem with body language more intense than in the beginning stages. As matters worsen, its groans deepen, it lies down, rolls violently and bites at itself. If allowed, it may chew holes in its body. And while rolling, it may twist an engorged intestine and die.

When pain of any kind persists long enough without relief, it naturally exhausts the horse. A horse debilitated in this way breaks out in a dreadful sweat, stands spraddled, ears flopped, head down, staggering to remain upright. It finally falls, legs stiff. It does not try to get up. Its life can be saved only if it is prevented from falling down, or if the downed animal is discovered in time to lift it to its feet and dull its pain with an injection.

With the pain of a snapped leg or other grave wound, some horses go into screaming panic violent enough to cause additional injury and death. Others lapse into shock, legs far apart, heads down almost to the ground, ears flopped to the sides, mouths open, eyes vacant.

The Sick Horse

A horse may be ill without feeling or communicating actual pain, or without frank symptoms such as puffy or weeping eyes, runny nose, abdominal noises, or the excretion of blood. Perhaps it has been sickened by contaminated food or water, or by drinking or eating too much or not enough, or by grazing on a poisonous plant such as oleander. Or it may have an organic ailment. Whatever the cause, you can tell that you are looking at a sick horse.

Mild Illness: The animal is listless, inactive, less responsive than usual. Its coat suddenly turns dull. It stands with lowered head, eyes half shut, ears in the airplane position (unless bodily pain turns them rearward).

Severe Illness: The sicker the horse is, the lower its head and the less responsive it becomes to persons, animals and events. Its attention is inward. It often groans. It trembles and sweats. It is uncomfortable whether standing or lying down, and repeatedly changes from one position to the other. In extremely severe conditions, periods of profuse sweat alternate with feverish dryness. When upright, the animal may stagger while struggling for balance on widely spread legs. At this stage its attention is virtually unreachable by human voice or touch.

In the earliest stage of illness, the most telling symptom is the coat. No matter how bright the sunlight, the dull coat of a sick horse reflects none of it. During recovery, continued dullness means that health has not yet returned.

The Hungry Horse

At meal time a normal horse expresses its eagerness for food in any of several ways. It paces its stall. It nickers. It kicks the door. It grabs the empty grain bucket in its teeth and rattles it as noisily as possible. If hay is coming, the horse probably is at its manger, rooting around, nickering, and occasionally nodding its head or even rolling it as in play. When the stall door opens and the hay arrives, the animal takes a great mouthful before the handler can deposit it in the rack. Most of the mouthful lands on the floor.

In stables where the horses get identical amounts of food at rigidly scheduled intervals, some remain hungry after finishing their portions. In fact, some behave as if famished. Finding not one grain or sprig of food on the floor or in the bucket, they set up a loud chorus of nickering accompanied by vehement wall-kicking.

When dangerously underfed, a horse looks the part. Its ribs show. Its dull coat seems mangy—coming out in patches, along with clumps of hair from mane and tail. Its posture is torpid, indwelling, with head down, ears flopped, eyes half closed. It is dying of starvation, yet has no eagerness for food. When nourishment is presented, it may raise its head and cock its ears, but it seldom eats without patient coaxing.

The Thirsty Horse

Whoever is responsible for the care of a horse should know beyond any possibility of doubt whether it has access to enough water. If the handler is not sure, the thirsty horse answers the question with considerable headshaking and licking of lips.

The ways in which they drink water vary with individual horses, perhaps depending on ancestral patterns. Some drink deliberately from the top of the bucket, without preliminary fuss. Others spend time splashing their mouths around the surface of the water before actually drinking it. This behavior may derive from eons of drinking at ponds where sweet, fresh water lay beneath congestions of algae.

Some horses dip their mouths a few inches into the bucket, leaving nostrils partly exposed. Others begin with the ritual splashing and then ram the nose against the bottom of the container, holding breath and chug-a-lugging from the bottom up. Bonnie had an eccentric Thoroughbred named Moose, who preferred to submerge his entire head in a barrel of water. When thirst was slaked, he would make a big sport of exhaling lots of bubbles, drenching bystanders in the spray.

After only a few days of thirst, a horse's body shows the ravages of dehydration. Deep cavities appear in the flanks, just forward of the hip bones. The normally smooth curvature of the abdomen from mid-barrel upward to the flanks suddenly becomes steeper and more abrupt, giving the animal a drawn or wasp-waisted look.

The Underweight Horse

Sooner or later, an overworked horse gets off its feed and loses weight. So does a worried, nervous, ill, injured or undernourished horse, or a mare in a difficult pregnancy.

The earliest loss of flesh is evident at the hindquarters, upper hind legs, flanks and chest. The point of the hip becomes abnormally prominent. Then the chest becomes flabby, as if the skin were too large for its contents. The tops of the ribs become visible at the front of the barrel near the withers. In extreme cases you see all the major skeletal bones.

The Overheated Horse

Newly arrived in a hot climate after months of adaptation to cold, or heavily blanketed in a warm barn after living comfortably in a pasture, or suddenly moved to a sunny pasture devoid of shade, a horse suffers acutely and quickly. It becomes lethargic. It dozes a lot. If asked for exertion, it washes out in overabundant sweat, flanks moving rapidly, nostrils flared, other bodily actions slow and sparing. It may slap its tail impatiently if a rider mounts.

The Cold Horse

Horses are regarded as cold-weather animals by persons who find it necessary to excuse or rationalize the common practice of asking the creatures for arduous work in subfreezing temperatures. The light-boned, warm-blooded horses most popular in the Western world are mainly of North African origin. That is, their ancestors seldom saw a day when the temperature fell below 50° F. If they have now become cold-weather animals, the laws of biological evolution have been overthrown.

To be sure, some horses tolerate cold-weather activity more readily than others. The others are recognizable. They shiver. Their muscles are tense. Their teeth chatter. But they lack the alertness of nervously quaking horses. Instead, they move reluctantly and slowly, responding sluggishly to their surroundings, ears dropped sideways or back on the head, eyes nearly shut. Their posture is typical of horses whose chief concern is internal.

The Lethargic Horse

We now describe conditions that range from the merely dull and dispirited to the barely ambulatory. The underlying cause may be any kind of debilitating sickness, stress or deprivation, either physical or psychological. Or it may be a constitutional defect, such as subnormal intelligence.

A horse in such a state responds sluggishly to external events. Something that might drive a normal horse up the wall provokes a subdued reaction in the lethargic one. The ears usually are sideways and move infrequently. The walking gait may include an actual shuffle that kicks up dust. If the animal lifts its feet higher, it drops them like clodhoppers. The head movements are sparse and unalert, eyes dull and heavy-lidded. In summer the coat is dull, in winter coarse and rough. The inactive tail may appear stringy or matted.

Nobody should expect a horse in this state to do anything well. In fact, physical abuse may convert lethargy to sourness without improving the quality of performance at all.

The Tired Horse

Sleeplessness: In a state of psychological upset or when quartered in a noisy place, a horse may be unable to lie down and sleep. It therefore dozes in an upright position or tries to. But dozing is not adequate rest. After a few days of this, the horse lapses into a dozing posture at times when one would expect alertness or even apprehensiveness. The phenomenon occurs behind the scenes at horse shows and occasionally is visible to those who can recognize it in the saddling enclosures of racetracks.

The sleepy animal stands head downward, eyes closed or almost so, ears in the airplane position—flopped to the sides with openings down. The stance is usually not square: One hip is dropped and most of the weight goes onto the opposite hind leg. The toe of the hind foot beneath the dropped hip points downward, bearing just enough weight for balance. At intervals, the horse shifts its weight. The pointed foot goes flat to the ground, its hip raises, the opposite hip drops and the toe beneath it points downward, barely touching ground. From time to time, the tail swishes or a muscle may quiver to dislodge a fly.

Overexertion: When a horse runs or does other hard work within the safe limits of its capacity, its ears prick toward the path ahead, or turn backward to the voice of rider or driver. Under heavy strain that is close to the threshold of physical crisis, the ears flatten from base to tip along the head. The nose reaches straight out, reddened nostrils flaring to receive maximum oxygen. Muscles bunch and, in thin-skinned breeds, veins swell conspicuously all over the body and particularly on the neck and around the eyes and face.

With the approach of exhaustion, the head "lugs" down in fatigue, alternately coming up for air and falling once more. The stride slows. When the mouth opens and gasping begins and the ears flop sideways, the animal is literally in a struggle to survive, having been pushed too far. At this point, with the horse's attention concentrated on taking in air rather than on work, the likelihood of a misstep and a snapped leg increases. And unless the animal is allowed to stop work at once, chances increase that its efforts will end with heart failure or internal hemorrhage.

Reined to a halt before the worst happens, the exhausted horse spraddles its forelegs for balance, nose almost to the ground, ears pinned back or flopping sideways, flanks heaving like bellows, breath whistling loudly. The body invariably drips with sweat. The base of the tail extends for inches, twitching spastically. If the handler tries to trot the animal off to the barn, the legs cannot raise the feet, which shuffle and stumble. Often enough, the animal collapses, lying flat out with eyes blank and glazed, legs stiffly twitching, all veins engorged.

The Herding Horse

To find greater physical security, or water or food or a salt lick, the lead mare sometimes herds the other mares and their foals from place to place. Her low head snakes from side to side, almost touching the ground on the extended neck. Her ears are pinned, tail somewhat raised and eyes directed threateningly at the laggard. She literally drives the others in the desired direction, reinforcing her authority by inflicting painful nips on the tender hocks and upper hind legs of those that retreat too slowly. The driving instinct is even keener among stallions. The unmistakable posture can be seen at its most highly developed among cutting horses exhibiting their skill at herding cattle. Magnificent photographs of wild stallions driving their harems away from competing males are included in Robert Vavra's remarkable *Such Is the*

Real Nature of Horses, published in 1979 by William Morrow and Company.

The Submissive Horse

The individual may not want to be herded away from one part of the pasture to another but obeys the higher authority and moves off peacefully. The act is one of submission, as is the flight of one stud from the combativenes of another. To appease the actual or potential wrath of older horses, weanlings and yearlings often demonstrate abject submissiveness by assuming a suckling posture, working their mouths like nursing foals. This seldom is seen in older horses except in extreme stress, as when a previously unraced three-year-old encounters a hard-boiled elder in the walking ring before a race, and surrenders by assuming the posture.

The Sexually Aroused Horse

Horses demonstrate libidinous feelings even when members of the opposite sex are not around. One mare may be aroused by the sight of another in heat. A young stud may be stimulated sexually by the nervous excitement that precedes competition.

Females: With intermittent squeals, the filly or mare stands head downward, hind legs spread, tail stiffened and held to one side, ears down at the sides of the head and open to the rear. Vaginal lips pulsate. Occasionally, the female signifies impatience by kicking with one or both hind legs and squealing.

Males: At the scent of a nearby female in heat, the stud lifts his head and curls his upper lip in the unique gesture of *flehmen*. He arches his neck, raises his tail and develops an erection. When he merely smells the urine of a mare in heat, he produces the *flehmen* and then, if allowed, marks the spot with urine of his own, possibly to stake a claim. The *flehmen* is not exclusively sexual but is the standard response to a strong odor of any kind.

It can be provoked by the scent of medicine, or perfume or even cigarette smoke.

The Drugged Horse

When intoxicated by a depressant such as a tranquilizing or pain-killing drug, a horse displays a singular pattern of body language. Quite different, but equally unique, is the body language of a horse supercharged by a chemical stimulant.

Unfortunately, the posture of the zonked-out show horse or racer does not reveal the brand name of the illegal potion that courses through its veins. All the expert observer knows is that the animal has had a little help from its friends or, contrariwise, has been medicated to lose. Knowledge of this kind is better than none. Race and show officials eager to encourage honest competition might well learn the equine language.

Depressants: Response to medication varies with the individual. The tranquilizing dosage that calms one show jumper and enables it to perform at its best might make a basket case of the animal's first cousin. And the chemical that merely elevates the mood of one horse may depress the physical reactions of another. As this implies, nobody can look at a group of horses and always identify every one that has been dosed. But one can recognize those on which dosage has had conspicuous effect.

The peculiar behavior of its ears demonstrates that a horse is under the strong influence of a depressant. As if the chemical had deadened muscles or nerves, the ears of an overly tranquilized horse do not function properly. They flop sideways in the airplane position, open toward the ground, and immobile. The animal may react to something in front of it, sniffing at it or even moving away from it, but the ears do not prick at it. They remain flopped. Except when its ears are pinched by ill-fitting blinkers, doping is the only situation in which a horse responds to stimulus quite readily with eyes and nose, while the ears behave as if disconnected from the sensory apparatus.

Under a less severe dosage, the ears may move in approximate coordination with the muzzle, but sluggishly. They never really prick forward, and they soon resume the flopped position. As this horse walks or jogs, the ears may actually flap up and down like those of a hound.

A secondary sign of depressant medication is a somewhat glazed eye and a comparatively sluggish response to visual stimulus. The head bobs low and, when walking, the feet shuffle and the tail moves slowly. A thin, dribbling drool may be noticed, suggesting inhibition of the animal's normal ability to swallow.

If severely affected by a depressant, a horse may act as if the chemical had suspended its sense of self-preservation. Occasionally, one sees a race or show horse walk straight into the rear of another horse that is kicking back at the intrusion. This is extreme behavior. Even blind horses do not walk into the kicking hooves of animals in front of them.

Stimulants: Again, the ears tell the story. They freeze in whatever position they may be in when the drug takes effect. They go rigid, as if the muscles were in spasm. Most often, they prick forward and remain so through thick and thin. The animal's other responses may seem approximately normal, but the stiff ears are hopelessly out of phase. For example, if the rider leans over to fuss with the bridle, the horse's ears remain stationary. Or if another horse gallops up from the rear, the overstimulated one's ears do not budge.

Depending on the degree of stimulation, all systems may become hyperactive. One sees an abnormal amount of head movement, erratic dancing and spasmodic muscular twitching. At the walk, trot or canter, the legs move with an unusually high action that sends the hooves flying at peculiar angles and brings them stomping to the ground as if directed at flies. Breathing is faster and less regular than normal. Even without heavy exercise, the flanks move rapidly and unrhythmically.

The eyes are probably glazed—"spaced out." Foam may gather around the bit and work itself into a thick, ropy drool that trickles to the knees. In males the base of the tail may stand out from the rump horizontally or upward, occasionally stiffening and bobbing in a kind of spasm.

A horse stimulated insufficiently to produce all the signs may seem at first glance to have tell-tale rigid ears. But then the ears may move a little. Keep watching. If the ears always return to the original position and if that position does not represent a normal response to ongoing stimuli, the animal is under the influence.

Overdoses: We have mentioned that the doper who administers a stimulant or depressant may discover that the substance has produced an effect opposite to the one intended. Just as in human beings, a moderate tranquilizer may drive a susceptible individual bananas. Or a mild stimulant may kill.

The equine victim of overdose looks the part. The eyes do not quite focus, but take on a wild, disoriented look. For no apparent reason other than the effect of the chemical, an overdosed horse may fall into totally inappropriate terror or rage, in which it can kill itself or its handler. Or it may freeze into a statue-like posture before keeling over.

The Buzzed Horse

When a horse backs straight into an electrically charged fence, the shock reverses the animal's direction, propelling it into a long forward leap. In that instant the tail stiffens at its base, swings in a violent full circle and then goes straight up in the air before slapping down to the haunches. All this happens in a second or two and is unlike any other equine response. It therefore is seen only near electric fences or at racetracks and other settings where equestrians use electrified riding crops to make horses move faster. Known as "gads," "machines," and "buzzers," and by numerous other jailhouse terms, the battery-powered whips have the energizing effect of cattle prods. For that reason they are outlawed in all forms of equine competition. For the same reason they remain in covert use by persons seeking unfair advantage.

If the racegoer ever sees a sudden leap with a circular swing of the tail followed by an immediate up-and-down slap, the only cause will have been a machine in the rider's hand.

The reader should bear in mind that the horse jumps directly away from the electric surprise. This means that its leap seldom carries it straight down the track. The leap is much more likely to be on a diagonal, inasmuch as the rider cannot zap the horse in the dead center of its rear end.

VOCAL LANGUAGE

Here we use the word "language" more loosely than before. Equine body language is precise. Equine sounds are not. Although they vary in pitch, loudness and timbre, the sounds seem mainly to be calls which emphasize the horse's need, wish or mood of the moment. The younger the horse, the higher the pitch of its voice.

Although scientists may some day establish that equine sounds hold specific meanings of their own, we and our readers are stuck with the lesser knowledge that now prevails. We cannot stretch that knowledge far enough to justify publication of a list of sounds and their supposed meanings.

For example, when a two-year-old colt enters a strange environment, it may whinny. One may choose to believe that it is saying, "I want my mommy." Or, "I dislike it here." Or, "I would rather be in Philadelphia." But all we really know is that horses utter distress signals in certain circumstances. To be sure that a sound is distressful, we need to appraise the circumstance as distressful, and we must make sure that the horse's body language conveys distress. Even having verified that the sound is associated with distress, we would have no license for anthropomorphisms like "I want my mommy."

In other words, glossaries of horse language do not work. We therefore shall content ourselves with an effort to describe the sounds horses make and some representative situations in which they are heard.

Snort: When this is not simply an effort to expel dust from the air passages, it may be heard at times of playfulness, impatience, apprehensiveness or annoyance.

Nicker: A generally nonaggressive sound, it is more nasal and not as loud or prolonged as the basic neigh. It often is heard in the call-and-response situation that exists when some horses are outside and their chums are in the barn. Or when horses greet each other or a human friend. Or when horses are away from home and, as happens in the saddling area and walking ring at a track, announce themselves to each other in a kind of roll call. Or when horses are getting hungry in their stalls. When Don Loudon entered his barns late at night, he always imitated a nicker to let the occupants know that it was only he and nothing to be startled about.

Whinny: A sound of higher pitch, associated with more tension and excitement than the nicker.

Neigh: To transmit over a longer distance than can be covered by a nicker, the horse lifts its head, opens its mouth and neighs. The broodmare neighs when her baby is out of sight. It replies with a whinny, not yet being able to produce a mature neigh. Or a horse that has heard no reply to its nicker may utter a neigh or two to attract attention from afar. Or a horse returning home from a lengthy absence will neigh when the barn appears on its horizon.

Bellow: Louder and deeper than a neigh, this authoritative sound is heard when one stallion announces its invincible presence, and some other stallion challenges that presumption of superiority with a bellow of its own. Angry mares and geldings utter that kind of sound as well. An even deeper abdominal call, yet not so raucous, it may be a perfectly friendly bellow for immediate attention. The horse that used to untie knots would endure the straightened rope for only a short while before slapping the stall door with its foot and bellowing for someone to come and make more knots.

Screams: These are immediately recognizable as such, although loudness and pitch differ with the vocal equipment of the individual and, of course, the urgency of the situation. Different screams occur during rage, fear or pain.

Squeal: Throaty, but an authentic squeal, this sound is heard among excited foals at play, or when a mare threatens to bite a pesky foal. It also is characteristic of a mare eager to be bred.

Sigh: When they lie down to rest, horses often emit a chesty sigh of contentment not unlike that of dogs or humans.

3. BODY LANGUAGE AND THE RIDER

Standing a few feet from either side of a horse, the reader of body language receives an entire message in one glance from nose to tail. A rider is less conveniently situated. From a normal position in the saddle, the rider of a forward-moving horse sees only the withers, some chest, the crest and a few additional inches of neck, the poll, the ears and some jaw. The position of the ears tells where the horse's attention is centered. By glancing at the shoulders the rider also can see which foreleg is on the ground and which is not.

And that is all there is to see. Even when the horse turns its head or the rider turns in the saddle, vision plays a minor part in the rider's comprehension of the mount's moods and needs. With most of the horse and its message out of the rider's line of sight, warnings of incipient trouble cannot be seen.

But some can be heard. And more can be felt. The familiar accolade that describes an expert equestrian as "part horse" is not hopelessly fanciful. Every good rider functions as part of the horse, with hearing and touch (especially) helping communications in ways that eyesight cannot.

Which is why many good teachers believe that the way to begin riding is bareback. Others recommend bareback riding after the student has learned the rudiments of mounted balance and coordination. The skin-to-skin contact of bareback riding makes significant communication of every tensing of human or equine muscle, every shift in human or equine weight. Sensitivity gained in that way is not lost when the saddle is reinstalled. Even with a 75-pound Western saddle between them, a good rider can feel a horse tensing its back muscles and arching its spine, preparing to buck.

To an enhanced sense of touch, an alerted sense of hearing and a limited range of vision, the good rider adds common horse sense. For example, in reaction to a loud noise from behind, the rider fully expects a horse to jump forward and then turn toward the noise to investigate. During the instant of the noise itself, the rider anticipates the animal's reaction and shifts weight to accept the forward leap without being unseated. Then the rider prepares to let the horse turn its head or even spin around bodily. To prevent the horse's natural reaction is to solicit a fight.

Another example: Riding in a group, you hear another horse approaching too closely. If your horse does not like that, you see his ears flatten, you hear his tail swish or slap and you may also feel his rump tighten for a kick. To avoid trouble, you immediately urge your horse forward and advise the rider behind to stop tailgating. You have only a second or two in which to act. If you are too slow, you may hit the dirt, and the horse behind may throw its own rider when trying to evade your horse's kick.

When your horse slows down and his ears are pricked forward, he has sensed something that he needs to check out. If you look directly between the ears toward their target, you may see what has caught the horse's attention. The matter of concern may be trivial but not to the horse. If his head drops and he wants to do a lot of backing and filling and snorting, it is time to dismount and lead him to the frightening object and let him discover, with your help, that it won't hurt him.

One particularly difficult maneuver for a saddle horse is backing up without turning the head to see the path behind him. The natural tendency is to turn the head sufficiently for rearward vision. But most good pleasure horses are trained not to do that and trust the rider. The rider assists the process by holding the reins lightly and issuing repeated assurances in a warmly confident tone of voice. The openings of the horse's ears should be directed backward in total attention. If they are not, the animal's focus is elsewhere, and chances are that the next move will be to jump out from under the rider.

The rider's alertness cannot be suspended even during periods of rest. It is fine to stop and chat with one's companions during a trail ride, but best to dismount before chatting. When engrossed in conversation with each other, human beings often turn off other goings-on, including the body language of their horses. If something startles one or more of the horses, the time for the rider to notice it is before the horse actually shies—that is, as soon as the ears prick forward, the head rises and the nose snorts. An oblivious rider is always in danger of a fall.

A horse impatient to go more quickly expresses this by ducking its head and then stretching its neck forward. The identical movement is typical of a horse whose mouth is being hurt by an overtight bit. The difference is that the eager horse's ears prick forward and it tries to move as rapidly as you will permit. But the horse who wants slacker reins slows down. His ears turn back to you and may also flatten. And you probably hear his tail slap impatiently against the hind legs.

It generally is easier to diagnose mild equine discomfort when mounted. From the ground you may notice that a horse seems to be bobbing its head a bit more emphatically than usual while walking. However, unless there is an obvious limp, you may be unable to tell which leg is being favored. In the saddle (or better yet, bareback), you will surely feel the hurting leg lift more rapidly, and you should also feel the extra tension and heavier impact on the opposite side.

When chafed by a saddle or girth or bothered by an internal pain, the horse shortens stride in hope of stopping. You can feel the body trying to bend away from the pain. The ears turn directly back to the site of discomfort.

Similarly, if the horse is overworked, you feel the stride shorten and the sides heave for breath. The head drops and the ears flatten. When the head turns, the flaring nostrils can be seen. The whistling breath is clearly audible.

4. SOLVING PROBLEMS

This happened at a justly renowned American institution of animal study. The prize stallion seemed to have come down with something. For more than a day, it had neither eaten nor slept nor consented to leave the rear of its stall. A small congregation of experts agreed that the animal was afflicted. But with what?

A passing student noticed the perplexity at the stall door. She found room, peered inside and saw her good friend the stallion quaking in a lather of sweat, sides heaving, eyes rolling.

"What's he afraid of?" she asked.

"He's not afraid. He's sick. We're trying to decide what's wrong."

"He's terrified!" she insisted. "That's fear, not illness."

"Could be right," shrugged one of the others. "But afraid of what?"

"Let me see if I can do anything with him," volunteered the student.

She got a lead rope and walked into the stall, talking re-assuringly to Friend Horse: "It's all right, feller. It's only me. I'm here to help."

They had worked together often. When she came close, he

pressed his head against her, shivering pathetically. She tried to comfort him. But when she offered to lead him out of the stall, he propped his forelegs and refused to go.

Tugging would not help. Talking might. She patted and begged. Would he please go out of there with her? After a while he gathered himself and sprang. In two bounds he was fifteen feet outside the door, dragging her behind him on the rope. He then wheeled and faced the door, backing and snorting.

"Notice how he jumped through the door?" she asked. "Now he's spun around to face the stall. Something's frightening him over there! What's new and unfamiliar?"

On the previous day, several bales of hay had been stacked next to the doorway of the stall. The horse's nose and ears were pointing at the hay.

"Let's move the hay and see what happens," suggested the girl.

When they moved the bottom bale, a large snake appeared. The horse reared and backed away another five feet.

Much surprise. "I'll be danged. Where did that come from?"

One of the men killed the snake with a shovel. When it lay still, the student asked the man not to remove it.

"This horse thinks his territory has been threatened. He needs to protect it. He's been scared for hours and he needs to vent his fear. Let him destroy the snake."

When she led the stallion toward the remains, his manner changed. His head came up. His ears pricked. His eyes flashed. With a loud snort and louder bellow, he reared high and stomped the snake to bits.

One of the men started to put the bales of hay back where they had been.

"No!" exclaimed the student. "Don't ever stack hay outside his stall again or he'll think there's a snake there."

Hay gone, particles of snake thoroughly sniffed, situation resolved, normality restored, the stallion ambled into the stall, tired but composed.

The story of the stallion and the snake indicates that knowledge of equine body language is not exactly rampant nowadays but is basic to the solution of equine problems. The challenge is seldom as unique as a snake in the hay. The solution rarely is so prompt. But the operating principles never vary:

1. *Recognition that a problem exists.* A handler who cultivates an interest in the horse's body language becomes considerably more attuned to the animal's feelings and is much more likely to recognize problems before they have serious effects.

2. *Identifying the general nature of the problem.* The body language seldom leaves room for much doubt: Fear? Anger? Boredom?

3. *Isolating the root cause of the problem.* This is not always possible. The difficulty may have arisen months or years earlier, when the horse was in other hands. Laborious investigation may be necessary. It will work only if the investigator knows the *kinds* of experiences most likely to precipitate the specific problem. That is, it helps (as always) to know something about the nature of horses, and it helps greatly to consider all problems from the horse's point of view.

4. *Treating the problem in the context of a sound working relationship.* This is basic. The stallion that had been scared by the snake already knew and trusted the helpful student. The *absence* of a good human-equine relationship usually is the chief reason that a horse's behavioral problems begin, persist and worsen. In any case, the problem is not constructively handled until somebody manages to earn the animal's trust. Having earned trust, which is no small feat and does not occur overnight, the handler must retain and nurture it. The prescription, once more, is patience and understanding.

5. *Never rush a horse into doing something that it is not ready and willing to try. When it balks at a new activity, retreat to the most recently acceptable one.* The horse needs a bit more time and preparation. This advice will seem preposterous to persons

who regard horses as unreasonable creatures that learn only under-
duress. Yet, as we have mentioned before and shall illustrate
repeatedly, normal horses of all breeds respond both reasonably
and enthusiastically to training that respects their instincts. By
suspending a new challenge until the horse is plainly ready for
it, the handler gets prompter, happier and longer-lasting results.
Persons unwilling to make brief, tactical concessions to horses
should not try to solve equine problems, being much more likely
to cause than solve them.

   6. *Never try to deal with a troubled horse when you are upset,
frightened, tense, angry, anxious or fatigued.* Horses pick up their
handlers' body language and other "vibes." A calm, matter-of-fact
handler gets best results. That handler knows how to relate to the
horse's curiosity and playfulness and knows also that the time to
stop work is before either party becomes bored, tired or irritable.

   To see how all this works, let us now consider the case of a
horse whose unrecognized and/or untreated fear has become
generalized, chronic and disabling.

The Frightened Horse

   When fear of a person or persons begins to dominate its life,
a horse probably slams itself into the farthest corner of its stall
when any human being approaches. To avoid any contact with
the dreaded species, the animal stands with head and eyes
averted until the intruder leaves. It even behaves this way toward
the person who brings food and water and cleans out the stall.
But look at its ears. They are constantly rotated toward the
visitor, scanning every move, revealing apprehensiveness. Nothing
is gained by attempting to touch this horse. In the beginning,
all effort centers on persuading the animal to leave its corner and
investigate you. This may take days but is indispensable. Until
the horse recognizes that you want to help, the situation is beyond
grasp.
   Bonnie's first step is to create a calm and interesting environ-

ment. She picks a time of day when activity is at a minimum and both human and vehicular traffic can be detoured from the horse's stall. She turns up there with a portable radio tuned to the schmaltzy background music in which many FM stations specialize. Horses love it. She posts herself on the other side of the breezeway, at least ten feet from the stall door. Why so far away? She calls it the horse's "safety space." She can tell by the horse's body language how far away she must be for the animal to feel secure. She sits on the ground and hums softly with the music. She does not talk.

The horse knows that something pleasant is going on, but is not about to turn around and look. Not yet. Perhaps not for days. To expedite matters Bonnie tempts his curiosity by doing macramé or oiling a rope or cleaning a saddle, remaining at the safe distance, leaving the radio on and humming in quiet accompaniment. Two hours of this each day, without other human presence, affect the horse deeply. The sounds and smells are irresistible. The horse's ears turn back more and more. He swings his head for little peeks.

Long after Bonnie has become sick and tired of macramé pot hangers, she arrives for the daily routine and notices that her patient does not fling himself to the rear of the stall. This time, he turns away quite slowly. Before this session ends, he turns completely toward the doorway, still safely in the corner but head forward, looking at her, ears flicking apprehensively. And then he walks a step or two toward the door, stops, gazes straight at her, ears pricked in real interest. She responds in kind, continuing to fuss with her strings or ropes or straps or whatever, yet talking directly to the horse, greeting him, congratulating him on becoming interested in the activities, describing in great detail the techniques of rope-oiling and generally engulfing the animal in quiet, friendly sound and intriguing sight.

He's hooked. A rudimentary relationship has started. A few more short, hesitant steps forward and he can peer outside the stall to make sure that it is safe to put his head out. That is, nobody is going to leap and grab him. And now at last his head is out, ears forward, safety space voluntarily reduced to five feet.

She can now move. She does so very slowly, still talking. She stands and offers him one of the ends of the macramé (or a thong or an oiled rope), reducing his space by the length of her arm but not stepping toward him.

"You want to smell this and see what it's all about?"

He stretches his neck, and she reaches as far as she can until his nose touches the object. Having exploited his natural curiosity, she now addresses his natural playfulness. When he takes the rope in his teeth, she lets it go. Folding her arms or keeping her hands quietly and unthreateningly at her sides, she stands her ground and watches the horse play with the new toy.

The animal has accepted some rope or string or leather from Bonnie but has not yet accepted Bonnie. To accomplish this, she tries to join the game. If she can reach the end of rope dangling from the horse's mouth, she may bat at it with a hand and make a big fuss over the result. By now, the horse is in the spirit of things, tossing his head in pleasure at the effects he achieves with the rope. She responds with laughter. Horses love laughter.

She holds the rope briefly, lets it go, grabs it again. Sooner or later, the horse bobs his nose up and down, looking directly at her, inviting her to resume play. Real contact has been made. She's in the game for fair, and well on the way to making friends. She narrows the space between them to three feet and switches to a two-foot rope. When the horse drops it, which is inevitable, she lets the rope lie where it falls. The game is now between her and the horse.

He reaches out with his nose, eyes friendly, ears forward in the classic get-acquainted gesture. She takes a step toward him and breathes gently into a nostril. He breathes back. She is welcome to step right to the doorway. He sniffs around her head and clothing. She scratches him under the chin and throat, being careful not to reach over his head, which might alarm him.

And now it is watermelon time. The rind of watermelon is a supreme delicacy. Just as judicious administration of champagne alters the disposition of a human being, watermelon rind beguiles the horse. In the present case, a horse that has spent weeks or months in a state of panic has been enticed into making friends with one person. The person now reaches into a pocket and produces a juicy piece of watermelon rind, strengthening the favorable impression already made.

Having achieved all that progress so rapidly, Bonnie believes it best to give the horse some time and space, lest the law of diminishing returns set in. She takes her macramé and departs, leaving the radio behind. An hour later she returns without macramé but with more watermelon rind. The horse probably comes to the stall door as soon as he sees her approaching. He undoubtedly checks carefully to make sure that she is alone, and then he extends his bobbing head with a low nicker, inviting her to come close. He gets his watermelon, some scratching and a good deal of friendly talk. She then leaves him alone.

The next day's activities begin as usual outside the stall but end inside. She has brought a soft body brush with which she grooms the horse thoroughly, brushing with one hand, scratching with the other, talking all the while in soft, encouraging tones. She is welcome in the stall. This is still a frightened horse but no longer frightened of this person. Every experience with this person has been a relief or a reward. Yet, when somebody else comes with food or water or to clean the stall, the horse still hides in the corner. Bonnie makes sure not to be there at such times. She also makes sure that the person who performs those duties is one with whom the animal has had no uniquely unpleasant experiences. Indeed, if she discovers that the horse fears one person more than others (suggesting that the problem may have originated there), that individual is no longer permitted within the horse's sight or hearing—even if this means excluding the horse's owner from his own barn.

With the beginning of routine grooming, Bonnie and the horse have become working partners. The emphasis shifts smoothly from watermelon treats to normal human-equine business. She takes over the feeding and watering, which means that the horse now accepts those necessities without crowding himself into a corner of the stall. To expand the routines and begin returning the horse to the world, she now appeals to his need for exercise—just as she had aroused his natural curiosity and playfulness.

She leads him around the stall on a rope. She then invites him to follow her through the open door. If he balks she does not persist, but returns to the stall and resumes the previous activity. She remembers the central principle: *Never rush a horse into doing something that it is not ready and willing to try. When it balks at a new activity, retreat to the latest acceptable one.*

Recognizing that the horse is not quite ready to leave the stall on a lead rope, Bonnie bides her time. She continues to mess around in the stall, leading the horse from wall to wall, grooming and scratching and encouraging. Soon, growing confidence and the basic need for exercise will encourage the horse to follow her outside.

Bonnie takes the horse wherever he can be turned loose for free play—a riding ring, paddock or field. She sits on the rail with the radio playing and lets the animal romp and roll, talking to him when he rushes by. The horse finally trots over for a bit of scratching, after which she pats him good-bye and leaves him alone in the field until feeding time.

After two or three days of these free-play sessions each morning and afternoon, the horse usually is ready for a walk on the road, the handler holding a lead rope but allowing the animal to browse. Another good activity at this stage is light work on a lunge line, allowing the horse to move in easy circles, changing gaits (if it understands the directions), backing up and otherwise behaving like a useful member of the outfit.

If the handler has been able to identify the person or activity most frightening to the horse, great care is taken to ensure the absence of that negative stimulus. By the same reasoning, if the problem is associated with weight under saddle, the handler does not yet attempt to ride the horse.

Sooner or later, something is sure to startle the horse. Perhaps a passing bird, a sudden clatter, a flying piece of paper. As soon as he shies or jumps, the handler gets very close, leans on him, pats and reassures him. The comforting body, hand and voice relax the animal. For all practical purposes (but we do not pretend to read the equine mind), the horse appears convinced that the person understands his fright. As the daily excursions increase, so do incidents of this kind. The handler always drops everything to provide comfort, and by the fifth or sixth recurrence, the animal needs less.

The horse has now become trusting enough to join the handler in bold investigation of frightening developments. For example, the handler has someone create a dreadful clatter with a bucket behind the barn. The horse may fall into a tizzy but quiets under reassurance and may be willing to walk around the barn to the offending bucket, sniff out its essential harmlessness and, in effect, learn that the handler is still dependably correct on all counts.

Success breeds success and confidence begets confidence. The handler may have been able to trace the animal's problem to some misadventure or series thereof with an individual person or object—Villain Number One. The horse is now ready to face and overcome the source of fear.

Perhaps it all started with a succession of fracases in a starting gate or trailer or van, or with an impatient rider. The horse has enough confidence in its handler to accept the confrontation, provided that the handler continues to observe the principle of never requiring the horse to exceed its own capacities. If it does not dare enter the van today, it will do so tomorrow or the day after. And it may not be willing to have the distrusted rider closer than ten yards away this afternoon, but will accept half that distance tomorrow. Horses learn. They learn especially well when allowed to do so at their own normal pace, without violence.

As usual, body language reveals the animal's attitude. Let us say that the feared human being has approached to within fifty feet, and, following instructions, is now standing still. The handler is grooming the horse. The horse's ears will prick toward the distant visitor. If one or both ears remain in that direction and an eye is cocked, the horse is apprehensive. If it turns skittish, the visitor should depart and try again on the next day. But if the horse soon redirects full attention to activities with the handler, the visitor is now tolerable—at fifty feet.

In due course, the horse remains calm and collected even when the feared human is seated only fifteen feet away and is engaged in lively discussion with the handler during the grooming routines. Soon the horse will permit the once-threatening person to come close, breathe into a nostril and begin working on the development of a totally new and much more constructive relationship.

We hurry to admit that this outcome is altogether beyond possibility if the human being has treated the horse with repeated cruelty. The only way to handle an irreparable situation between horse and master is to find the horse a new home and give it a new start. But if the problem began with an impatient mistake, complicated by a failure to recognize and treat the effects of the mistake, recovery is probable. For example, a man was upset by something unrelated to equitation. While performing in a cutting class, he had flown off the handle and whipped his horse so unexpectedly that they both fell. The horse was terrified of the man for weeks. But after treatment of the kind prescribed here, the pair are now reunited in a friendlier relationship than before.

And now some hints about the relief of fears attributable to traumatic experiences not necessarily associated with human cruelty or impatience.

The Head-shy Horse: The animal has a fit when someone tries to put a bridle or halter on its head or even touch its head at all. Perhaps a thoughtless rider once tied the horse's reins to a fence rail. Something startled the animal, it jerked its head and the bit tore its mouth. Or the problem could have begun when it banged its skull on the low doorway to a shipping van. As usual, the guiding principles apply.

To solve the problem, the handler must win the horse's trust and must retain it by practicing patience—never pushing the animal beyond its ready capacity. In time the horse accepts this person's hands at its head, having previously accepted grooming and scratching and general friendliness elsewhere on its body. Getting the tack onto the head is now a small detail. Perhaps the handler puts molasses on the bit. Or, if the horse is especially sensitive about the ears, the handler may unhook the straps, get the bit into the mouth and then hook the bridle behind the ears. Or the horse may accept the halter when it can reach the coveted watermelon on the ground only by putting its head through the formerly dreaded apparatus.

The Water-shy Horse: Maybe it slipped and fell when a rider forced it through a mud puddle. Or a groom may accidentally have struck an eye or ear while hosing the horse. Need we repeat that correction of the first problem begins with the establishment of a trusting relationship and proceeds stepwise until the day when the animal is ready to approach a mud puddle? As for the horse hurt by a hose, there is no good reason to expose it to that particular fear. Sponges and pails are preferable.

The Saddle-shy Horse: Is he actually afraid of the saddle, or does the problem begin whenever someone opens a large piece of cloth such as a saddle blanket or a cooler? To expedite acceptance of the cloth (*after* winning acceptance of your more fundamental attentions), it helps to impregnate the material with the horse's own scent. One easy way is to rub some of his manure into it. If the difficulty concerns the saddle itself, leave the animal's back alone until he is willing to accept activity there. Then begin by applying a small, lightweight cloth to the saddle area. Be careful that nothing dangles. When the time comes to try an actual saddle, remove the girth and stirrups so that the burden can be removed instantly if the horse demands. And begin with a small English saddle rather than a cumbersome Western one. After the saddle is accepted, add a five-pound bag of flour. And so forth.

The Angry Horse

The first horse mentioned in the preface to this book was a good jumper whose body language proclaimed acute annoyance.

The animal's rider was an awkward, heavy-handed young man whom the horse finally dumped.

Now it happens that the rider was also the owner of the horse. This means that the rider had inflicted his clumsy reinsmanship on the horse's tender mouth many times in the past. The horse undoubtedly had demonstrated irritation more than once. But nothing constructive had been done to prevent the magnified annoyance that left the rider sprawling in the tanbark of a show ring.

How could he have avoided that occurrence? Let us consider the situation in light of the operating principles previously recommended.

1. *Recognition that a problem exists.* If the owner-rider or someone in his entourage had understood equine body language, the animal's irritation would have attracted notice when first expressed.

2. *Identifying the general nature of the problem.* Clearly the annoyance was related to the horse's work—jumping.

3. *Isolating the root cause of the problem.* The horse was considerably more adept than its rider. The rider was hurting the horse's mouth and restraining its head in positions that made jumping difficult. It was reasonable to deduce that the animal's complaint was directed at the rider.

4. *Treating the problem in the context of a sound working relationship.* Easier said than done. The natural tendency of many equestrians is to blame the horse rather than the manner in which it has been handled. This young owner, while agreeing that the horse is angry at him, might decide that the animal should be taught to be less irritable and more accepting. But the only way to achieve those goals, without resort to force, is to earn the horse's acceptance by riding more competently. Force might indeed curb the horse's reactions and spare the clumsy rider some

falls. But force would not enable the horse to take jumps when its head was being held in the wrong position. The rider would be no better off than before and would have a dispirited horse in the bargain.

The prescription, then, is to remove the stiff-wristed rider from the horse. Let the young man improve his equitation. In the meantime let the horse be ridden by someone with better hands. If the owner cannot become a better rider, he should sell his spirited horse and obtain a more phlegmatic one. Which reminds us that the proud parents of juvenile equestrians should obtain expert advice before allowing themselves to be talked into buying the impetuous youngsters larger and more spirited horses than they really can handle.

5. *Never rush a horse into doing something that it is not ready and willing to try. When it balks at a new activity, retreat to the most recently acceptable one.* If our irritated jumper has been upset too frequently by its owner-rider, its grievance may no longer center on the young man himself. For example, it may become angry even when approached with a saddle—recognizing the equipment as part of a pattern of distressing experience.

The effort to help such a horse differs in only one essential from the program that overhauls a frightened horse. If the animal harbors anger of a generalized character, its temper can flare with little advance warning. None but an experienced horseperson of supreme alertness should risk an approach. Even a mildly irritable horse is dangerous to a less than fully capable handler. Until the horse recovers from its problem, the handler keeps an eye peeled at all times for the flattened ears, slightly raised hind hoof and tensed hindquarters that precede attack.

Otherwise, the handler concentrates as usual on earning the horse's trust. Recognizing that an angry horse needs no additional irritants, the handler is quick to remove them—or else remove the horse from them. For example, if the horse's problems are compounded by biting flies, it is important to limit their depredations. Call the exterminator. Move the animal to a less fly-ridden place. Rub the horse with fly spray twice a day. If the insects bother its eyes, get a fly guard. When the horse shakes its head, the dangling leather strips brush the flies away from its eyes. Although the horse will never write an essay or deliver a speech about your kindness, each measure you take in its behalf contributes to its awareness of you as a helpful person.

We mentioned that the horse might go into a snit at the sight of a saddle. Keep saddles away from it. Later, after the relationship has grown, it will accept a saddlecloth on its back and then, with little objection, a brief experimental period under actual saddle. For all intents and purposes, you are now home free.

Or the horse may have become angry at being confined in a stall. Keep it out of the stall at all costs. The day will come when it consents to enter the stall for a meal.

Perhaps we should emphasize that the purpose of the program is not to stop an animal from being angry at unreasonable or incompetent treatment. The purpose is to eliminate anger by eliminating its causes. To return to our example of the show jumper and its inept owner, let us now assume that the young man was learning better technique while the horse was being tended by softer hands. Their reunion would be cordial enough. As soon as the horse found that the rider no longer inflicted pain, everything would be fine. Given half a chance, horses are forgiving creatures.

A horse that the beginner should avoid is the one that is aggressively angry at the entire human race. Not many places keep horses of that kind. Yet even they are sometimes responsive to proper treatment, as we shall show in Chapter Six.

And now for some hints that may facilitate dealings with an angry horse.

Investigation. If the reader sets out to determine how a chronically cranky horse got that way, previous owners or handlers may be unable to answer the question. As already noted, it is exceedingly difficult for a human being to accept responsibility for a horse's misbehavior. The reader may find, however, that an hour or so around the horse's previous barn is enough to solve the mystery. When you see how the folks treat other horses, you get clues to the effect they may have had on yours.

Venting Anger. Despite best efforts, it rarely is possible to insulate an irritable horse from all causes of anger. When it becomes angry, the handler removes the cause as quickly as possible—or removes the horse. And then the handler makes sure that the horse discharges all its anger as rapidly and completely as possible. The best way is to turn it loose in a paddock and let it settle itself. This is far different and far more effective than punishment. Even when violence makes a horse more tractable, it enlarges the animal's grievances and diminishes its will to perform.

The Sour Horse

In 1977, when he was three, Silver Series won the Hawthorne Derby, Ohio Derby and American Derby in a period of twenty-eight days, defeating good Thoroughbreds like Run Dusty Run, Cormorant, Jatski and Affiliate. Later in the season, he ran second to the immortal Forego in the Woodward Handicap.

At four the gray colt won the Widener Handicap and lost twice by the margin of a nose to Run Dusty Run in the Seminole and Hialeah Challenge Cup handicaps. He was an extraordinary horse. And for a while he was the sourest good horse that anyone in Florida racing could remember having seen. He was balky, reluctant, resentful and fractious. He was a bundle of grudges, negative to a fault. He disliked everybody and everything. He hated to go to the track for his morning exercise. Exercise over, he hated to return to the barn. He would pin his ears and buck. He would try to run off. On days when tourists were invited to have breakfast at trackside and watch the horses work out, the management asked Silver Series' people to keep him in the barn. Management did not want the public to see a good horse battling its handlers.

In the spring the sour Silver Series stopped exerting himself in races. He was beaten by animals that should have been no match for him. But William duPont III recognized that the colt had great potential at stud. Furthermore, it was by no means too late to reconfirm his standing as a first-rate runner. With all that in mind, duPont paid $1.5 million for a controlling interest in the colt and turned him over to Peter M. Howe for training. Howe already knew that Silver Series was a thoroughly sour horse. He also knew that the problem undoubtedly was a reaction to hard racing, hard work between races, harsh punishment and stern routines—never idle, never free.

Howe tried to gentle Silver Series in his own barn, but it was too late. The colt remained negative. One awful afternoon at Saratoga, he threw a fit and actually sat down in the paddock—by way of expressing reluctance to run in a race.

That was that. Howe took Silver Series to Montpelier, the Virginia farm of Mrs. Marion duPont Scott, and turned him loose in a paddock. "He walked to the dead center of the paddock, stood there quietly looking around for a while and then rolled on the ground," remembers Howe. "Then he got up, tossed his head up and around and rolled again. We kept him off any kind of track for weeks. Sometimes we rode him in the fields, up- and downhill."

The next stop was a Kentucky farm of William duPont III. There he was ridden by Kaye Bell, herself a trainer and former race rider. "All I did was gallop him in open country, letting him do what he wanted," she says.

When the Kentucky track, Keeneland, opened its race meeting in October, Silver Series had not had a single workout since August. And he was a new horse. He won three races on successive Saturdays, including the Fayette Handicap, in which he equalled the Keeneland track record for a mile and one-sixteenth.

"Between races, we had him back at the farm," says Peter Howe. "We played with him and let him graze."

This was unorthodox treatment for a Thoroughbred racer, but it was problem-solving of the highest order. Howe recognized that nothing could be done with Silver Series unless the colt were removed from the stall, liberated from routine and allowed to refresh his spirit in natural surroundings. That Silver Series responded with three successive victories and a record-equalling performance should provoke thought in racing circles.

The Perverse Horse

Healthy horses are exceptionally playful. When not absorbed in feeding, sleeping or working, they seek amusement and are quite inventive about converting humdrum situations into games. We have shown that lack of opportunity for diversion may lead a horse to boredom and from there to destructive attitudes. Similarly, when a horse becomes playful at an inappropriate time or in an inappropriate way, the negative reaction of an irritated handler may provoke a patterned contrariness. For the perverse horse, the object of the game is to drive the handler cuckoo.

The behavior resembles that of a child who teases to see how far it can go. We have already met the horse that entertained itself by refusing to let the trainer slip a bridle onto its head. Another familiar type takes a step or two sideways, perhaps by pure coincidence, when the saddler reaches to put the saddle on. When the handler fusses at this inconvenience, the horse pursues the matter, eliciting huge reactions with small maneuvers. After the saddle is finally in place, the same horse may well dance sideways whenever the handler tries to put a foot in a stirrup and climb aboard.

Another well-known pastime of perverse horses is deep inhalation just as the handler tightens the girth. After the animal exhales, the victim discovers that the girth is not tight at all. When quartered in a field, this type of horse amuses itself at human expense by breaking down a particular part of the fence as often as the management rebuilds it. The horse may stand there and solemnly watch the hired hand repair the damage. On that person's next service call, the horse stands less solemnly and enjoys the spectacle of a human throwing a fit.

So long as the human continues to take the bait and becomes upset at the horse's pranks, the animal enjoys the sport, regards the person with the cordiality usually extended to playmates, and may indeed be a respectful and responsive workhorse in other situations involving the same person. On the other hand, if the person refuses to bite and reacts without fuss, the game is not worth the horse's trouble and is soon abandoned.

Of all reactions to equine teasing, the least constructive is anger. When that emotion expresses itself in physical abuse it discourages playfulness, to be sure. But it also discourages the intelligence, liveliness and interest that playfulness represents. Worse, it directs the relationship toward any of several destinations far less desirable than teasing. Frightened, angry or sour horses do not get that way all by themselves.

If the handler has recognized that the animal's perversity is merely inappropriate play and has responded without excitement, but the horse persists in teasing, some other way must be found to spoil the game. A calm order to "Quit" may restore decorum when accompanied by a swat to the ribs. This hurts neither the animal's ribs nor its dignity, but directs its attention to business.

Another useful tactic is the invention of a substitute pastime. For example, instead of repairing the fence in the usual way, add one loose board to be kicked around. Or equip the pasture with a surefire plaything like a sack of tin cans.

By now it probably goes without saying that the handler's knowledge of equine body language ensures that playful teasing will not be confused with hostile fractiousness. The horse with flattened ears and whistling tail is not playful.

The Bereaved Horse

Its restless pacing, lengthy nickers and persistent peering into the distance indicate that the animal misses someone or something. The handler surely knows the specifics of the situation and whether the separation will be permanent.

Even when the separation is brief, a grieving horse benefits from extra food and special human attention. Otherwise it may fret itself into a dull-coated bag of bones. Sometimes high-potency grain and extra attention do not suffice. Bonnie remembers a horse that went into an alarming decline when its master was hospitalized. Both spruced up after the hospital granted permission to have the horse led onto the grounds for a through-a-window visit with the delighted owner.

If the departed companion does not return, the horse may recover from its grief in a few days or weeks or months, or never. One determinant is the profundity of the horse's need for that particular companion. Another is the quality of the environment in which the bereaved animal lives. One quite representative pattern is seen on perfectly nice farms with excellent facilities. The lonesome horse finally stops nickering for its missing friend but becomes a loner, holding itself aloof from the other horses and not even joining their pasture romps.

A loner eats and performs adequately but without real zest. Its body is likely to be gaunt, its coat slightly dull. Its spirit will not return until it finds a friend or the management supplies one. The arrangement had better be more enduring this time. Two grievous separations can cause permanent damage, leaving the horse mentally and physically depressed.

A horse that had barely recovered from the severe effects of a traumatic separation was purchased by the parents of a fifteen-year-old girl. The stable manager sat down with the youngster and said, "Please do not become that horse's buddy unless you are sure that you can continue. Another separation will harm him."

The girl was not certain that she could find the time to maintain a close relationship with the animal. "Okay," said the manager. "Just be his friend. Be nice to him, but don't let him get the idea that he's your main interest and that he can count on you to spend hours with him on a regular basis."

The girl's first enthusiasm dwindled in a few weeks. She visited infrequently. Her horse never achieved the ultimate sparkle in the coat or the extra twenty-five pounds of flesh that it needed. When the parents offered to sell the horse, the manager knew of a likely buyer, a nearby summer camp. The camp horses lived on the range during the off-season. One of them was a gray gelding that had never fully recovered from the death of its closest equine friend. The manager hoped that these two lonely ones would hit it off. They did. Within two weeks they were romping through the hills like a couple of colts and looked healthier than they had in months.

The Sleepless Horse

Without careful medical examination, it may be difficult to tell whether a listless horse is ill or simply needs sleep. In either plight the animal may go off its feed, lose weight and spend an unusual amount of time in the dozing posture, with head down and ears flopped sideways. If the hind feet and hips shift position from time to time, the animal is surely dozing. Constant day-time drowsiness means that it is either sick or sleepless. But if the ears frequently rise and turn toward some part of the body, the horse is injured or ill.

The owner of one especially sleepless horse lamented that it had formerly been a creature of rambunctiously playful spirit but had suddenly turned languid. It spent most of its day dozing. The veterinarian had found no infection, no sign of organic distress, no injury of any kind.

The owner assured Bonnie that the horse had not been over-worked. She replied that something must have changed to prevent the animal from lying down and sleeping at night. But the owner could provide no clues. No equine chum of the sleepless horse had left the barn. No new horse had come to the barn and upset the animal's priorities. No new dogs. No new fence. No change in airplane traffic or other nocturnal disturbances.

"Wait a minute," said Bonnie. "What about that railroad track out there? It goes right past the barn. Any new trains going by at night?"

The track was seldom used, said the horse's owner, but he seemed to recall hearing something about a newly scheduled freight coming through at night.

"Through" was the right word. A little detective work found that a freight had begun to pass the horse's stall six nights a week at precisely 4 A.M. The stall was situated so that the engine's powerful headlight shone directly on the poor horse, shocking it awake. The prolonged clatter and the glare of the train's running lights were the finishing touches. Four in the morning was torture time. The horse was becoming a nervous wreck, and understandably so.

Bonnie and the owner could not change the railroad's timetable or move its tracks. Neither could they move the horse: The owner had no other stall for it.

"Won't he get used to the noise?" asked the owner.

"Certainly," said Bonnie. "But he will never get used to that light. Perhaps you can do some minor construction and keep the light out of the stall."

The work was completed before nightfall. Within a week, the horse had recovered its appetite and bloom of health and had stopped dozing all day, having resumed the practice of lying down for a good night's sleep.

The Convalescent Horse

The mental consequences of injury may heal more slowly than damaged tissue does. Until the horse is mentally ready to resume work, it cannot be regarded as "recovered" from its accident.

We have touched on this kind of problem in earlier pages. One good example was the filly who had been confined to her stall for a week and absolutely refused to exercise under saddle when pronounced physically able to leave the stall and return to the track. After confinement of that kind, a horse may not only resist work but may fight desperately to avoid returning to the stall when work is finished. This is particularly noticeable if the horse has been cooped up for weeks, as happens when it is held upright in a sling while waiting for a broken bone to heal. The animal emerges from the ordeal with a true case of cabin fever, the intensity of which diminishes somewhat when the handler has sense enough to transfer its living quarters from the hated place of recent confinement to a new stall in a different barn, or at least in a different part of the same barn.

A horse injured when cast in its stall may then fear all stalls.

One that bangs its head when rearing in a stall may fear all places with low ceilings. If either horse is to recover its former quality, the handler must exercise considerable patience. The first step, as usual, is to recognize that there is a problem. The next steps, also as usual, are to identify the problem and its cause. The horse must be removed from the cause of the problem. Which means that it must be turned out in a paddock or pasture and must not be transported in a roofed trailer. Then, laboriously but with every reason to expect success, the handler must go through the entire process of transforming a frightened horse into a confident and willing one.

In some cases, mental rehabilitation is impossible or impracticable. An old racer had suffered bowed tendons three times and twice had recovered sufficiently to return to action. After the third bow healed, he again came back to the track. He seemed as willingly cooperative as ever, except that he did not run swiftly in his workouts and always returned limping to the barn. As soon as he entered the barn area, he stopped limping. He did not limp when walking around the breezeway or when hitched to the carousel exerciser. He did not limp when walking to the track for his work. The veterinarian could find no sign of a new bow. There was no heat in any part of any leg. Apparently the horse continued to associate the track with remembered pain. He limped after exercise because he expected to limp after exercise. This cannot be proved, of course. Nor can the alternative theory be proved that the horse limped in hope of persuading his owners to let him quit racing. In any event, they finally retired him because they could find no way of getting him to run without limping.

The Vices

Several kinds of unapproved equine behavior are known as "vices."

Cribbing. This is known also as crib-biting. The bored horse learns to amuse itself by gnawing the stall door, the manger or any other wood that is accessible. Some handlers combat this with restraining straps, which supply frustration without solving the central problem. The cribbing does the animal's teeth no good, but the main problem is the swallowing of air—"wind-sucking." This bloats the stomach, interferes with digestion and leads in time to lung damage.

The best time to deal with this problem is before it becomes established. A bored horse needs more work, more diversion, more play, more fatigue. Pending the beneficial effects of a more constructive routine, all cribbable objects should be removed from the stall. The top of the gate should be kept closed, covering the favorite gnawing ledge. Many cribbers are diverted by the presence of a volley ball on the floor of the stall.

Weaving. The horse passes time by shifting its weight repeatedly from one foreleg to the other, usually with head and neck swinging from side to side. If boredom is the cause, the horse needs to be out of the stall as much as possible. If it can be moved to a field, so much the better. If not, it should be put into the stall only to eat and sleep. Some weavers develop the habit because they have sore front feet or knees. These should be checked by a veterinarian as soon as the behavior appears.

Tail-rubbing. When bothered by an itchy anus associated with worm infestation or incomplete grooming, a horse may develop the habit of rubbing its hindquarters against stall walls. In time the upper tail and the flesh at the point of the hip may be rubbed raw. If the habit persists in an animal that is properly groomed, free of worms and adequately exercised, it may be necessary to install a low shelf on all the walls of the stall. When the horse tries to rub its tail, the shelf will catch it at thigh level and prevent contact between tail and wall.

Head-rubbing. Horses sometimes feel an overpowering desire to rub their heads on something. This may begin with the itch of dried sweat, or neglectful grooming, or noses and eyes congested by dust or colds. Rubbing creates problems for humans and other horses, especially when it becomes habitual. The equine head is so ponderous that an attempt to rub it against the body of another horse is an act just short of assault. And it can push a human being off balance. One solution is to hang an outdoor chaise mattress on an accessible part of a fence and let the rubber rub away. A less delicate alternative may serve occasionally. The handler faces a wall, bends toward it from the waist, propping the hands on it at something like arm's length, allowing the horse to rub its itchy head on a human posterior.

Kicking: Some horses, especially barn-sour ones, try to kick their handlers. An especially effective countermeasure involves a four-foot length of old-fashioned rubber (not plastic) hose. As the horse begins to kick backward, step sideways and whack it on both hocks. The loud, hollow "thunk" of impact is so startling that the horse spins around to face it and you. Few repetitions are needed and the hose inflicts no real pain. In some circumstances the handler teaches the lesson by kicking back—sharply planting the toe of the boot in the horse's abdomen. This does not hurt the animal as much as it would hurt a human being, but it certainly captures its attention. Beginners should not try to treat this vice. Others should do so only if sufficiently alert and nimble. When caught off guard by a kicker, the only hope is to move toward the horse, colliding with its hocks rather than its hooves.

Stall-kicking: A horse with an irritated hock discovers the relief of bumping against stall walls. So may other sore-legged horses. It does not take long for them to become full-fledged wall-kickers. In some cases kicking denotes boredom, hunger or anger. In others it is only a habit. But it can knock shoes loose, chip hooves and wrench muscles and joints. It should be nipped in the bud with whatever assortment of work and diversion can keep the animal contructively occupied.

Biting: Young horses are mouthy. They inspect things by nibbling at them. If allowed they inspect human beings that way. Even when babies, they can bite forcefully enough to break skin. Later, they can break bones. The handler should wear no attractions like shiny buttons or fuzzy sweaters. To discourage biting, poke a fingernail lightly into the soft side of the youngster's muzzle. As you do so, say "Quit!" in a low, firm voice. Note here that "Quit" is always preferable to "No," which sounds too much like "Whoa" and causes no end of confusion between handlers and horses. After a few applications of the fingernail, you will need only to point the finger to remind the baby not to bite at you. Nevertheless, even a well-trained three- or four-year-old colt may resume nipping as an expression of studdishness. Again, the fingernail works. If the horse goes further, actually charging with open mouth, the handler can effect rapid cure by tossing some red pepper directly into its face. Or, if the animal enjoys nipping the hand that holds its lead rope, it will change its mind after a few short squirts on its lips with a pocket-size breath-freshening spray. The horse will hate the sound, the smell and the taste.

Important Safety Measure: It must be evident by now that an equine vice is an objectionable or inconvenient behavior pattern that derives from conflict between the nature of the horse and the unnatural circumstances of domestication. The problem with which we will now deal is not regarded as a vice but is more dangerous than most and, as usual, brings worst harm to persons unprepared for it. When you take a horse to its beloved pasture after protracted imprisonment in a stall, you should expect exuberance and take pains to protect yourself against its hazards. Lead the horse through the gate into the field and then turn it around so that it faces outward toward the gate. You then turn it loose and *step backward* through the gate, keeping your eyes on the animal. It probably will rear a bit, pivot on its hind legs, kick rearward in jubilation and race away. If you follow any different procedure in turning it loose, its hooves may level you.

Confusion

A major obstacle to swift correction of problems is the horse's inability to understand why the handler demands that it stop doing something previously acceptable. The older the horse and the problem, the greater the obstacle.

Let us say that its previous owner enjoyed letting a horse snatch a handkerchief from his back pocket. The nippy little game went on for years, amusing both parties and becoming a hallowed tradition of their relationship. But now the horse has moved into the barn of a new owner, who has not been briefed about the handkerchief game and, worse, is unacquainted with the body language of a playful horse. After being nipped in the rear a few times, the new owner (who carries no handkerchief in his back pocket), hauls off and hits the horse for its apparent misbehavior.

The horse becomes confused and resentful. If the new owner does not complain to the previous one, find out about the game and learn to keep his behind out of harm's way, things are sure to go from bad to worse. On the other hand, if apprised of the game, a patient handler might be able to substitute some other pastime.

Another horse may have spent months learning to prance. The new owner does not enjoy it, preferring that activity be limited to the walk, the trot and the canter. Not realizing that the prancing was learned behavior, the uneasy owner might misunderstand it as fractiousness and might even be afraid that the horse was preparing to bolt. If the person becomes tense and punitive, the horse falls into utter confusion. It is quite probable than an experienced handler would recognize that the horse was otherwise tractable and that the prancing was a mere affectation enjoyed by the previous owner. The habit would then be less menacing and could be modified and perhaps eradicated by patient encouragement of other gaits.

Total success is not always possible. Some behavior patterns are too deeply entrenched. One familiar example is the fear that prevents an older horse from entering a van or a covered trailer. A patient handler may finally be able to coax the animal into an open trailer, especially by accompanying it on the journey. But it may never willingly enter a van.

In Appendix A, we shall explain how to avoid buying serious problems when buying a horse. It is no fun to discover that one's new purchase, so winsome in the seller's riding ring, suffers from a problem that affects its work, limits its charm, and raises serious questions about the advisability of continuing to feed and house it. Truly dominant problems may be beyond correction, but usually not for the horseperson who (a) knows how to deal with them and (b) has the time.

In Chapter Six, we shall tell about the rehabilitation of a so-called "killer" horse, an out-and-out rogue that was unwilling to cooperate with anybody. The process began, of course, with the establishment of a working relationship, which was no small feat in itself. And then the horse had to be reeducated, as if it were a newborn foal.

The realities and practicalities of horse ownership being what they are, not many readers will have occasion to school foals or retrain killer rogues. Nevertheless, we think that the principles and practices discussed in the next two chapters will serve well in more casual dealings with horses of all kinds.

5. THE EDUCATION OF
    A FOAL

The schooling of a horse begins at birth, when it learns to respond to the instructional touch and voice of its dam. In many places, human beings involve themselves in the care of the newborn foal so that it will accept their authority as perfectly natural.

That principle is valid enough, although some persons try to apply it without due regard for the role of the dam. To assure oneself of a willing, friendly foal, it is imperative to earn the acceptance of its dam. This can hardly be done instantaneously on the day of foaling. Superior results are obtained by befriending the mare many weeks before she foals.

Whether human beings handle her foal or not, the mare will be the primary influence in its life. It will learn quickly from her and be slow to forget the lessons. It will emulate her behavior and absorb her attitudes. The would-be schooler of a foal should spend lots of time with the pregnant mare, grooming and feeding and exercising her in adequate space. In some breeding establishments, all this may be difficult to arrange without administrative upheaval. Broodmares get minimal attention in many places.

Let us assume that you have arranged to purchase the unborn

foal of a mare quartered on a good farm. You have become her friend. She accepts your presence in the foaling stall when she gives birth. You become part of the foal's first experience, helping to clean him, talking to him quietly, helping him balance himself and manage the first suckle. You do not linger overlong. The two need privacy and space. But you visit frequently during the first three days to fuss gently and quietly with the baby. You always talk to him. You let him lean on you while he tests his wobbly legs. He becomes accustomed to your touch, your voice and your scent.

In most places mare and foal go to pasture three days after birth. Now that the surroundings are more natural, you can begin to press the advantages that you negotiated for yourself during the final weeks of pregnancy. You begin by doubling the attentions that you paid the mare before the foaling. If you formerly spent one hour a day feeding and grooming her, you now spend up to two. The curious baby will be fascinated by the grooming implements and what you do with them. In a while he will also be enthralled with the special feed that you bring the mother. He will accept a coffee can of the sweet feed or nutritional supplement or whatever it is you bring, and will grow more rapidly with the extra nourishment.

Because the foal finds you mighty interesting, he is all over you like a puppy. Whenever he displays an interest, satisfy it. Show him the cloth or comb and use it on him a little. Calmly, of course. And quietly. When he does something amusing, laugh in appreciation. It is not at all unusual for a week-old foal to solicit human giggles, recognizing them as kinds of compliments. For instance, the baby is sure to imitate the mother's grazing posture, as if he were interested in grass. When you laugh at him, he probably enjoys the reaction so much that he promptly assumes the spraddled position and earns the laugh on your next few visits.

Or he may tangle his legs and fall in a heap. You help him to his feet, chuckling a bit, and say, "That's all right, little guy. Now let's get you together. Put this foot over here and we'll take you back to Momma to settle you down."

In one sense the foal understands not a single word you have said. In a more practical sense, he has understood every syllable as a sign of friendly helpfulness from a creature worthy of trust.

So long as you remember that the relationship all began as a friendship between you and the mare, and so long as you continue to treat her as a friend, you remain a welcome part of the family. Some mares become jealous if a human being—even a friend—ignores them in favor of their foals. Some become alarmed if a human detains a foal for even a few seconds after the mare has called it. Do not monopolize the baby. Watch the mare's body language carefully. At the first sign of irritation, go to her, settle her and leave.

If you get into a squabble with the mare, you create a grievous problem, the solution of which may take many days. The foal's attitude will reflect the change. If you are unable to appease the mare, you may never succeed in winning the full confidence of the foal.

As if all this were not sufficient to establish the delicate character of human-mare-foal relationships, we must now mention the skittish nature of the foals themselves. They are jumpy little creatures. They react to surprises and other sudden changes in exaggerated ways. Their spirits are preposterously high. Their fears are quickly kindled. Their attention spans are infinitesimal except, as we shall see, when they are playing games. They are handfuls. They are babies.

It goes without saying that the human being who punishes a suckling foal for being a baby immediately loses the friendship of the dam. Neither will the foal's mama feel much more cordial toward a person who makes the baby unhappy by frightening or irritating it. To prevent even a hint of such discord, you should be guided by principles not much different from the effective problem-solving strategies discussed in the previous chapter.

1. Emphasize play. It is the very best way for the youngster to work off excess energy and calm itself for attention to lessons. And those lessons must always be games enjoyed equally by teacher and pupil.

2. Some unpredictable mishap can occur at any time. Immediately stop whatever you have been doing and return an excited foal to its dam.

3. When introducing the foal to a new challenge, such as a new skill, be careful never to force the issue. At the slightest sign of reluctance, retreat to the previous successful activity. Your forbearance makes the foal feel better about you and helps assure greater receptivity the next time you introduce the new maneuver.

4. Emphasize the positive not only to the foal but to yourself. At this stage of its tender life, you are much the larger and stronger. He understands at once that you can control him. He also understands that you are gentle. And he loves to provoke your laughter, your appreciation. Later, even if he stands sixteen hands and weighs half a ton, he will retain this basic respect and affection for two-legged, two-handed, talking creatures who treat him decently.

The Halter and First Restraints

On the theory that a foal should regard tack as a natural part of its bodily equipment, many establishments adorn the new babies with halters, which become permanent equipment. For handlers employed in mass-production, with large numbers of foals to "break," the ever-present halters are a convenience. The controlled environments of those outdoor factories are usually free of the low-hanging branches and broken fences on which a haltered baby might hang itself. In other circumstances, the halter is more a danger than a convenience. Furthermore, it deprives baby and handler of useful experiences together.

Where the emphasis is on individual schooling, you bring two halters to the pasture, one for the mare and the other for the young one. He watches as you put the halter on his mother. He sees that she accepts it. By now he is well accustomed to your touch on his head and body. You show him his halter. He mouths it with interest. You merely slide it over his head and hook it. When you leave, you unhook and remove it. Daily repetition prepares the youngster for later realities such as restraint.

On one of the first times the halter is applied, a clumsy handler may frighten the foal by being too forceful about keeping his head still. A battle may be disastrous. With a handler clinging to the halter of a baby trying to run off, a fall is inevitable. The youngster may never recover from the distrust and apprehensiveness caused by such an experience. On the other hand, if you are careful to immobilize the head for only a split second, the foal soon submits to more prolonged restraint—especially if you invariably let him go as soon as he expresses the wish.

By the age of two weeks, the foal accepts not only the halter but a lead rope. The first experience with the lead comes after you have let him watch somebody else attach one to the halter of his dam and walk her around him. You then hook up his own lead rope. In another two weeks, he will enjoy being led on it, with his mother alongside on another.

The foal's accommodation to restraint should not be limited to the head, of course. His natural itchiness helps in this respect. Welcoming your scratches, he happily allows you to lean all over him, briefly preventing him from moving in any direction. As he grows and begins to shed his fuzzy foal fur, he becomes even itchier and therefore even more amenable to restraint while being scratched. If he moves away from your touch, you are scratching too hard.

And then there are the feet. During the baby's first day or two of life, he surely permits you to pick up and reposition one or more of his feet to help his balance. Later, when you school him to lift his feet for grooming, he will not be afraid of falling down, having learned to associate attentions of that kind with improvement of balance.

All of this should happen easily and quickly, so long as you continue to honor the priorities of the dam. She is still boss. You do not approach the foal when he is resting or feeding, or when the mare obviously wants him beside her or when her body language conveys the least displeasure. Always capitulate to her. She may be mistakenly displeased, but she is the dam and can turn the youngster against you.

You will find morning and evening the most opportune visiting periods, when the mare is feeding or looking for her food and the youngster is most playful. Never initiate play yourself. Approach slowly. When the baby senses that his mother sanctions a romp, he tenders the invitation with a toss of the head, a flip of the tail and a little buck. You never decline because, for the young horse, play is an ideal prelude to work (which is itself a form of play). So crouch and slap your thighs and leap about and wave your arms gently, and try to respond to the bounding baby's body language with easy gestures of your own, plus appreciative laughter. Like a romp with a large dog, this is great exercise for both parties and strengthens the bond. But be careful: Foal play includes rearing, wheeling and kicking. The kick of a two-week foal can shatter human bone.

The Lessons of Grooming

Our object is to spare both the animal and its handler the traumatic crises associated with traditional "breaking." Why should horse and human have to fight with each other over something as inconsequential as a saddle when the whole bother can be avoided by thoughtful grooming during the early weeks of life?

After the natural transition from fingernail scratching to the use of currycomb and body brush, you introduce a soft, white rubbing cloth. The baby accepts this in his ears and mouth, over his eyes and everywhere on his body. You can wave it in front of his eyes without alarming him. You can even flick it all over him. He actually likes that. His acceptance eliminates the likelihood that he later may be taken aback or even terrified by aspects of bathing or of medical treatment. The flapping cloth also is good psychological preparation for the day when a saddlecloth will be put on his back.

Even better preparation for saddle comes with the next addition to the grooming equipment, a burlap gunnysack. It comes into play as soon as the foal has become fully accustomed to the white cloth. He will find the rougher burlap pleasurable when wadded and used as a body cloth. Now you begin unfolding it. Let it rest on his neck or rump or back, flapping in the breeze. In a few days he will actually nap with the gunnysack over his ears and eyes. He even will walk on the lead rope with the sack on his neck or head.

The gunnysack provides a great game. Throw it at the foal's head. He tries to catch it in his mouth. When he does so, he throws his head around, wrapping it in the sack. You then unwind the sack from his head, scratching him as you go.

Another kind of gunnysack play teaches him to become accustomed to dangling things. He learns that no harm comes from stepping on the end of a gunnysack dangling from his back or neck. Sometimes you soak the sack in water and accustom him to feeling the wet cloth over his body and in all its orifices. The first time somebody soaps him up for a bath, he will accept it without turning a hair. But that is merely one of your accomplishments at this point in the young life of your foal. Look:

1. In his later youth when you sling a saddle pad on his back for the first time, he will not be frightened. No big deal for a youngster that has been through so much good training with a gunnysack.

2. He will not be disconcerted by the feeling of a body blanket, thanks again to the gunnysack. Furthermore, if part of his loosened body blanket hits the floor of his stall some night, he will be able to step on it without becoming frantic, having gone through the drill with the burlap.

3. When fire breaks out in a barn, the horse is trapped in its stall. A human rescuer arrives and tries to put a wet gunnysack on the horse's head. If the horse fights this, its life may be lost. But pasture games taught your horse to accept a wet gunnysack on its eyes and face.

We have already noted that your early attention to the foal's feet taught him to associate such activity with improved balance rather than with a terrifying loss of balance. At one month the youngster is ready for additional lessons. He probably lifts either forefoot cooperatively as soon as you touch it. When he does, reach around, grab the pastern and restrain the foot for a split second. Increase the period of restraint in easy stages, avoiding alarm. If the foal does not lift the forefoot automatically when you approach it, run your hand down the rear of the leg, pressing slightly. The hoof will rise reflexively. In a few days you will be able to hold the foot long enough to groom it properly. But always let go the instant that the foal becomes nervous and throws his head up.

As you run your hand down the back of his leg, always lean against the foal's shoulder, obliging him to bear his weight on the other three feet. He must learn not to lean on the person who grooms his feet. And, lest we forget, he also must learn not to nibble at the person's shirt or jeans. The first time he does that, try pushing his head away with the flat of your hand on the side of his muzzle. If that does not work, a gentle poke with a fingernail will get results.

Those foot and balance exercises and the lessons in grooming etiquette will spare the young horse later difficulty. Horses that do not know how to behave during grooming sometimes provoke severe retaliation by irritated handlers. To give your foal the best possible opportunity to reach his potential, you must do everything possible to protect him against brawls of that kind.

Serious Work

With the dam's consent (and often to her great relief), you have been leading her frisky foal farther away from her until most of your grooming and schooling exercises take place as far as fifty feet away. But not out of sight. At two months the foal has learned an astonishing repertoire of skills. He walks beside you without jumping around or leaning on you. He stops when you say, "Whoa." In response to gentle urging with the rope, he turns in any direction. He also backs up.

The secret of this success is his eagerness to play these games well and earn your praise. In walking on the lead, if he leans against you, you lean back, forcing him onto his own feet, and praise him when he walks well. When he starts jumping around, you stop the parade, settle him and refuse to go on until he is still. For this obedience he earns accolades.

You teach him a left turn by standing at his left shoulder, pulling on the lead rope with your left hand, and pushing his rear end away with your right hand. He walks a full circle around you as you pivot in its center. He learns the right turn with you at his right shoulder, pulling at the lead rope with your right hand, pushing his hindquarters with your left. Later, he will have no trouble performing the same maneuvers on a thirty-foot line and, perhaps more amazingly, without any line at all, responding entirely to vocal commands and hand signals.

You teach him to back up by tugging the rope straight back so that his nose bobs toward his chest. Your other hand presses on the chest as you say, "Back!"

Which prepares him to learn how to reverse direction at the the command, "Reverse!" To the usual pull on the lead rope and push on the rump you now add a backward step which turns the youngster completely around. In a surprisingly short time, he will associate "Reverse" with this vigorous maneuver and, perhaps more surprisingly, will eventually perform the reverse on the lunge line when you utter the word and swing your arm in whatever direction you would push his rump if you were working at close quarters.

In this process, and all by itself, the foal learns the slide stop, plus the spinning and pivoting that are so important to nimble breeds. He is still a suckling baby but is thoroughly familiar with the lead rope. He can walk with it dragging and can step on it without fear. You can move it all over him. For example, you can rest it on his withers. When you cluck at him, he follows you as if under rein—but you need not even touch the rope. Naturally, this may take a few days to learn and may begin with a certain amount of discombobulation. If so, abandon the project for the time being, withdrawing to a more familiar activity at which the foal is sure to succeed. At this stage such an activity might be a sequence of walk-trot-whoa-back-whoa, all on the lead rope with vocal instructions and little or no hand pressure.

He will finally master the leadless stroll and be ready for the ground tie, the first really unnatural demand you will make. When you drop the lead rope to the ground, he will learn to accept that as a signal to stay put. He will learn it as a game. The two of you are head to head. You drop the rope. You step back two paces. When he steps forward to follow, you raise your hand, touch his chest and say, "Whoa. Stay put." When he stops, you say, "Atta boy," which he cherishes. You will have to repeat this, with many false starts and no little confusion, until he gets it straight. Bend over to face him at his own eye level for best communication. Take your time. Finally, he will stand still when you are eight feet away, your hand still upraised. Do not overdo it. After a maximum of fifteen seconds, drop your hand, praise him and call him to you.

Before you know it (and we mean in a matter of days), you will be able to retreat as much as fifty feet from him as he remains loyally ground-tied. He does it to please you. He would rather be next to you. He has learned the first of the unnatural accomplishments that will make him a good horse.

Dividend: In the process you have taught him to come when you call.

You will now find it quite easy to teach the foal to stand still as you stretch the lead rope to its full length, walk all around him and drape it over his back. When he starts to pivot to keep you in full view, your familiar "Uh-uh, whoa" and upraised hand instruct him to halt. When he resists all temptation and follows instructions to the letter, your relationship has been strengthened considerably. Also you have prepared him for his later work on a lunge line (known in some places as a long line).

At about three months—and in any event before he is weaned —he should have his first experiences with a trailer. It should be a regular, roofed-box trailer—the kind that terrifies some horses, perhaps because it arouses instinctive fears of dark caves.

Lead him and his dam to the trailer. Let him sniff all around it. Open the side door, revealing the small amount of sweet feed that you had put on the floor. When he pokes his nose in and nibbles, you practically take the words right out of his mouth: "Hey, there's good stuff inside that trailer!" End of first visit.

On subsequent visits the sweet feed is always there. So is the dam, bored and showing it. And you take to drumming on the metal sides of the van with your fingernails, and drumming on it with your open hand, and banging fenders, tires and wheels with your feet. The baby checks your body language and that of its dam and sees no reason for alarm. Apparently a little clatter is part of this particular deal, along with sweet feed.

At perhaps the fourth visit—if all the previous ones have been successful—have somebody hold the mare and foal while you walk up the loading ramp into the trailer. Open the side door, stick your head out at the baby. "Boo! I see you!" or any congenial greeting goes over very big. So does a handful of sweet feed. Bonnie remembers a baby so enchanted with this part of the game that he broke away from the handler, rushed to the rear of the trailer, dashed up the ramp and joined her inside. His dam got a bit worried and was led into the trailer to get some goodies of her own.

A more usual introduction to the trailer takes somewhat longer than that. You lead mare and foal up the ramp at the same time, having taken care to remove the center partition from the interior and having had the vehicle scrubbed clean to eradicate any stallion scents which, by disturbing the mare, might provoke anxiety in the foal. It may be needless to point out that you went through the trailer routine with the mare before she foaled to make sure that she had no problems with it.

As the young one accompanies the mare into the van, you stand on his side of the ramp, between him and the edge, with a hand at his halter. The lead rope has been removed. If a foot slips, you are there to help. As he passes, your hands touch his side. As soon as he is inside the van, you go to the side door and join him, handing up some sweet feed or a slice of watermelon rind, keeping him busy. His mama is now engaged with a pound of sweet feed, which will last about five minutes, during which she transmits no messages other than those of contentment.

Your little foal has now spent five full minutes in a trailer or van, which is plenty. It is time to get him out and properly. In later life he will have to back out of vans. He therefore should back out of this one. Unload the mare first but do not allow the youngster to turn around. Back him out. It will be a delicate operation, considering the absence of the mare and the unsure footing of the ramp. But the foal trusts you. He has already done things willingly for you that would never occur to him in other circumstances. And he already knows how to back up. With a little planning, the unloading will go smoothly.

You hook up his lead rope as the mare backs away from him and down the ramp. He probably nickers for her and turns his head. Rather than let him turn around, busy him with the well-known backing-up game, right down the center of the ramp with your reassuring hand on his rump and your voice telling him how well he is doing. Take him all the way to the bottom. No jumping off the side. When he reaches the ground, let him go to the mare and nurse. He surely will want to after the adventure. When he finishes, fuss all over him for being such a wonderful fellow. Take the pair of them back to the pasture and turn them loose.

After four drills of this kind, the baby will back out of the trailer unassisted, even before his mother does. Moreover, he will tolerate the van or trailer with the back door closed and the engine running.

He goes from triumph to triumph. His next is an actual trailer ride around the barn. You and the mare are right there with him. She is perfectly unconcerned, enjoying her sweet feed. He never gets the chance to regard this or any other experience as overwhelming. As soon as you see the body language of apprehensiveness, you back up. You discontinue whatever arouses the fear. You return to a more acceptable activity.

Before this foal has reached the age of four months, you can leave him and his dam alone in the trailer while you drive them a few miles down the road and back.

The Big Separation

The foal loves each of the educational routines. Daily repetition of the entire assortment might be tiresome for you if each day were not a challenge to the young one's own interest, intelligence and affection. Every day you add something new, yet you never neglect to run through all skills previously acquired. He lifts his feet for grooming and restraint. He turns. He waits. He comes at your call. He incites you to romps. He plays with the white cloth and the gunnysack. He does his stint at the trailer at least once a week.

At some point in his fourth, fifth or sixth month, he will be weaned. Deferred weaning gives him fuller advantage of the maternal influence. The time of weaning is usually dictated by human convenience, and is seldom delayed so long that the mare attends to the matter herself. By whatever reasoning the date may be determined, you begin preparing at least a month in advance. Indeed, preparations actually begin a few weeks after the foal's birth, when the mother so willingly allows you to take him yards away from her. You now do more of that kind of thing—always in her sight, to be sure, but farther and farther away and for as long as both will tolerate. You lead him off to explore. If he gets interested in a flower, you let him sniff it and taste it. Then you pick it and tickle his nose with it, which starts a general frolic.

You take them both from the pasture to a barn. You hook the mare to a hitching post and let him go exploring with you, always in sight of her, but in ever-widening circles. If he is interested in rocks, rock-rolling is a splendid pastime. He may get lost in the game but not altogether. When she nickers, he nickers back. If that does not satisfy her and she calls more urgently, you bring him to her at once.

He is sometimes reluctant to return, but rules are rules. He now wants as many side trips as he can wangle. If anything frightens him, you are there at his shoulder, your bodies lightly touching. And he can jump against you for help as if you were his dam. She would then lean across him and quiet him with her voice. In most minor upsets, you can do almost as well with verbal assurances and your arm over his neck. If he is severely rattled, you quickly get him to the mare, who will probably have noticed the trouble and begun calling. But when his body language and hers express nothing urgent, you pat him and say, "It's nothing. Let's check it out." You get the piece of paper (a frequent offender), and ball it up, and let him sniff it and lip it and get over it.

All the routines that were initiated in the pasture are now fully accepted in front of the barn. You run through them there every day. One day, when it is time for the mare's grain, you walk them both into the stall. After inspecting everything in the enclosure, the foal eats calmly alongside the mare. As a veteran of trailers, he has no hesitancy about accompanying his mother into a harmless stall.

On the next day, you do not hitch her to a post but put both of them directly into the stall. You then take him out to play where she can see him, after which you groom him in the breezeway right in front of her stall. Now comes the ever-expanding program with the lead rope, an activity that takes you strolling farther and farther away until you finally are out of the mare's sight. If the foal nickers or she bellows, you hurry him back to her.

You have been increasing his food, so that nursing gets less emphasis. His own sense of independence grows rapidly when you not only remove him from her field of vision but turn him loose in the exercise ring for completely free play. He can see you and can come to you if he likes, but may choose to entertain himself for many minutes before doing so.

The mare has not complained of his absences for many days, and they have become longer each day. When separation comes, it is virtually painless. You take the dam to another farm—or at least out of earshot. She goes into a stall for two or three days until her milk dries. During that period, frequent human companionship will help. Meanwhile, the youngster is either in a stall and paddock or turned out to pasture with other weanlings or an older horse or both. He may have a few rough moments, but not serious ones. In a few days he is fully detached from the mare and contentedly attached to you. The weaning is over. His dam returns to the broodmare pasture and life goes on.

Rapid Advancement

You place a pole on the ground. He walks over it. He steps on it and lets it slide under his foot. This may alarm him. When you assure him that he is safe, he believes your tone of voice. You have never swindled him. If the pole bothers him more than that, making him fear a fall, you follow your basic principles—retreating to the previous successful activity. Tomorrow will be time enough to try the pole again and learn that one can slide a foot around on it without falling.

Now come four poles, two end-to-end on either side of him, forming a lane through which he walks or trots with his head down to see properly. And with a long, five-foot-wide lane of poles, he will trot through, back through, walk halfway, turn and trot out at your order.

You introduce a mud puddle. He inspects it carefully, finally extending his neck to taste the water. To demonstrate that puddles are things to walk through, you jump in. He then experiments with a front foot and decides to take your word for it. He soon walks and trots through that mud puddle without turning a hair. If he is a Thoroughbred, Standardbred or Quarter-horse racer, he has just taken his first lesson in moving through slop. Meanwhile, you both are splattered from head to foot, but you have added another to your lengthening list of astonishingly easy accomplishments.

Whether he is to be a race horse or not, you may want the foal to feel at home in wet weather. Let him walk through the lane of poles during rainfall or under a sprinkler. In your playtime tone of voice say, "We're playing in the rain! Fun!" Both that foal's parents and all his brothers, sisters, cousins and aunts may stand head down in rain, doing as little as possible, but this one quickly learns to enjoy its possibilities.

At some stage, either before or after the introduction of the poles, you start him walking through rubber tires. You show him one. You sit on it for him. You roll on it, demonstrating how harmless it is. "Come on. Step in this thing." The foal goes through the usual ceremony of inspection and complies, if not on this day then surely on the next. When you put four tires next to each other, he learns rapidly how to walk through them, watching carefully to place his feet on the ground rather than on the rubber. And you are alongside, giving him assurance and security.

Back to the poles. You raise one of them about six inches off the ground. "Okay. Nice and easy. Let's go over this thing." With you on the lead rope right at his shoulder, he steps gingerly over it. A career as a jumping horse may not be implied in his pedigree and may indeed be no part of your plans for the foal. But this is marvelous exercise. His coordination increases daily as he learns how to watch where he is going and do whatever is necessary to get there.

Take four 15-foot poles and make an L-shaped path of them, four or five feet wide. Get him walking the path without touching the poles even while making the turn. When that becomes old hat, back him through it. You naturally are there guiding him with your hands on his withers and rump. When his foot touches a pole, you say, "Whoa. Now move up a little bit and start again." He knows what you want of him. He tries his best. He succeeds. He becomes beautifully limber, bending in the middle to place his feet correctly at the corner.

Now you make it more difficult. If the poles were four or five feet apart, place them three feet apart. When he now succeeds in negotiating that turn backward or forward, he will have conquered one of the testing maneuvers required of show horses entered in Western pleasure classes. The help you gave with your guiding hands at withers and rump before he took over the whole procedure himself was a foretaste of the reining and foot and leg pressures that riders will use later.

If you feel like it, you can have him go through rubber tires placed inside the L-shaped path of poles while playing a gentle hose on him. You probably will feel like it and so will he because everything is still a game, just as from the first day of your working relationship.

And now he is all of six months old, a prize weanling more independent, confident and adaptable than others of his tender age, and able to do things beyond the abilities of many five-year-olds. Yet his schooling program seldom takes more than an hour a day. Time now to increase that.

Lines and Reins

As a big shot of six whole months, the foal is able to tolerate more work and absorb new knowledge at a faster rate. You will find it worth while to spend two hours a day at his lessons. Continue to begin the morning with a thorough romp. If he is in a weanling pasture, he will have companions to play with and should do so. Then you and he stalk each other, play tag or whatever other kind of mutual foolery has become the custom. About fifteen minutes of this reconfirms the playful basis of the relationship and works off his excess energy. Then it is grooming time, including all the well-liked routines with cloths and feet, after which comes work itself. Much of the two hours will be spent exercising all previously acquired skills, and the rest will be given to new ones. But be sure to leave plenty of time for the post-work rolling and resting which precede lunch.

He is ready for the lunge line, a thirty-foot length of one-inch nylon or cotton webbing. You use it instead of the lead rope, attaching it to his halter or to the muzzle ring on a similar contraption called a cavesson. The line rolls up easily in your hand, so that you can put the youngster through exactly the same whoa-walk-trot-reverse-whoa maneuvers that you long have been doing with the lead rope. But now you can lengthen the distance between you and the foal and let him strut his stuff ten feet away. You are too far away to touch him. He no longer feels your comforting hand. Chances are large, however, that he readily accepts this change. After all, you are still visible, still involved, still talking and gesturing.

Within fifteen minutes, he is the full thirty feet away from you. By the third of these sessions, he may already be responding unerringly to your hand gestures and vocal commands, changing gaits, stopping, starting and reversing without any change in the tension of the line. You shorten the line gradually and when you are close, you put him through, "Walk! Whoa! Trot! Whoa!" and then, without preparation, "Back!" You raise the directing hand, and he probably backs up all by himself, just as he has been doing on the lead rope for weeks.

After two months of this, you turn up one day without the lunge line. You hook the lead rope to his halter and begin the drills at close quarters, unhooking the lead rope almost as soon as you begin. He will do all the routines entirely unrestrained, responding to your gestures and words, enlarging the circle all the way to the outside fence thirty feet away from you, prancing with his proud tail up, having a fine time.

He is ready to feel the sensation of reining. You begin with long reins, which are essentially two 15-foot lunge lines. One attaches to the left side of the halter, the other to the right. He wears no bridle or bit. The truth is that he never really will need a bridle or bit as long as he lives.

You exert the lightest possible tension on these long reins. He already knows how to turn on command. You now supply the faintest, most delicate rein pressure to support the vocal instructions. For the first few days you do all this at the youngster's shoulder. After he has become accustomed to turning left or right while feeling the corresponding pull of the rein on his halter, you can stand directly behind him at a distance of perhaps five feet, cluck to him and direct him to walk. He will surely turn his head to see what this is all about, but he will then move out. You can now "drive" him in either direction, make him stop, let him trot briefly. Because you are taller than he is, and your voice and rein pressure come from behind and above, the feeling for him is much like that of carrying a rider—except, of course, that he now bears no weight. And allow us the indulgence of repeating a previous promise: A horse trained this way will *never* need a bit.

You are preparing him for saddle. Obtain a surcingle—a belt

sometimes used over a saddle and girth to keep the equipment in place. During the grooming one day, put the surcingle on the youngster. Tighten it just enough to stay put. Do not insist on leaving it in place if he objects, as he probably will. If he tolerates the band around his body, so much the better. But be sure to remove it as soon as he indicates that he has had enough. After he becomes used to the surcingle, you can thread your long reins through its rings, elevating those lines to a realistic height. At the same time, you can apply a light saddle pad (a bareback pad), which weighs only about one pound and will not impress the youngster as more onerous than his familiar wet gunnysack.

After you achieve total acceptance of the surcingle and pad, replace the surcingle with a light training saddle, which will lack stirrups but will be fitted with ring attachments for the long reins. In a week the horse is schooled to a stirrupless saddle and mighty responsive to reins.

Although he does not require a bridle and bit, he very probably will get them sooner or later and had better be introduced to the equipment by you. The bridle part is easy. It resembles a halter so closely that it can be substituted without disturbance. But the bit is a foreign object in the mouth. If at all possible, avoid this part of the schooling until the youngster is at least eighteen months old.

Begin with the lightest rubber Pelham bit you can find. Help yourself and the horse by smearing molasses all over it and your hand. Let the bridle dangle from the bit as you bring it to his mouth. As he tastes molasses and begins licking your hand, slip the bit over his tongue and into the mouth. Let him spit it out as soon as he wants to. But during the second or two between entry and ejection, if you have managed to slide the bit all the way into his mouth, press it gently against the corners. The first time you try this drill, prolong it for not more than two attempts. Try about three minutes the second time. On perhaps the fourth or fifth day, you may be able to slip the bridle on before he spits out the bit. This is sure to upset him. He will be working the bit with his tongue and lips, trying unsuccessfully to find a comfortable position and finally attempting to expel it. Let him do so and do not wait more than fifteen seconds before slipping the bridle off with the bit. On later attempts you can wait half a minute.

The relationship between you and the horse is such that he will accept bridle and bit. The more adept you are at managing the tack (and the youngster's mouth and ears), the more readily he will adjust to this new demand. You can now work him under saddle with the bridle and long reins. Remember at all costs that messages between you and the youngster have always been delivered vocally or visually. Do not misuse the reins. That is, do not overuse them. The lightest possible touch will do. The bridle is actually redundant. The horse can do his work without it. And remember, it is best not to spoil things by pressuring his mouth with a bit. Wait until he is two years old if you can.

Which brings us to the question of weight on the back. Unless he is to be a race horse, he should carry no substantial weight until he is three years old, and then only for brief periods. At two, three and even at four, equine bones are still soft and growing. The orthopedic problems that terminate the careers of young race horses are mainly attributable to premature weight-bearing at high speeds.

A properly trained yearling like yours can carry a ten-pound sack of grain (anchored to the saddle) without being harmed or flustered. He is already accustomed to the sensation because you have been leaning across him throughout his life. Until he is several months older than two years, he should not carry more than fifty or sixty pounds for more than a few minutes. As he approaches three, you may lie briefly across his back, lift your feet off the ground and quickly jump off. When you are ready to sit on him, lie across his back in the familiar way and then pivot on your abdomen until you can sit astride.

All of them react identically to this experience if they have been sensibly prepared for it. They turn their heads backward with a perplexed look, unaccustomed to having you up there.

"It's all right, fella," you say. Then you cluck and order, "Walk."

Off he walks. So much for "breaking" a horse.

No horse educated in this way need ever feel a rider's kick, or the prick of a spur, or the sting of a whip, or a sharp jerk of the reins. He is a supple creature, acutely responsive to the slightest leg pressure, shift of weight or touch of rein, as well as your vocal instructions. If you now teach him the basic movements of dressage, you will prepare him for whatever career specialty you consider suitable, from jumping to cattle-cutting. The sporting libraries of the world contain abundant technical information about dressage and other forms of higher education. We need not rehash that material here. Instead, let us enter territory in which traffic is sparse. Let us tell how the principles of problem-solving and basic schooling may affect the behavior of a rogue horse.

6. THE REHABILITATION OF A ROGUE

When his undersized, thirteen-year-old son expressed interest in becoming a jockey, the man decided to buy the kid a race horse. In their section of southern California, practically every backyard had a barn and exercise ring. The man would put a Thoroughbred back there and let the boy try his hand.

A few telephone calls produced a fellow who had a horse that was unable to race anymore but was plenty of animal and could be bought cheap. Why couldn't he race anymore? Lame? Nope. Sound as a dollar. To tell the truth, the horse couldn't race anymore because the stewards at the track had banned him. The thing of it was that this horse hated racing. Kicked up fusses in starting gates. Threw a few riders. But only when asked to race. Otherwise, this horse was a real pussycat. You could practically make a house pet of him.

While driving to her parents' home in that California neighborhood one day, Bonnie heard screams. She stopped the car and rushed to the exercise ring behind the man's house. The voices were horse and human. She saw a huge Thoroughbred with blood streaming from its chest. It reared high and brought its front hooves crashing down on the shoulders of a prostrate boy. The

155

boy's father stabbed frantically at the animal with a pitchfork, drawing more blood.

Bonnie was able to reach beneath the bottom rail of the paddock fence and drag the boy to safety. Both his shoulders were smashed and he was unconscious. She noticed that the father had also escaped from the enclosure. The screaming horse slammed into the fence, trying to continue the attack. Bonnie heard the wood crack and saw the rage in the animal's eye. She noticed blood coming from his mouth as well as his chest.

To make the horse more manageable for its young rider, the man had equipped it with a Spanish spade bit. Few bits are more cruelly punishing. Bonnie found later that this one had lacerated the horse's tongue and torn the roof of its mouth. Apparently the horse had gone berserk as soon as the boy's first yank on the reins tore flesh.

The father gasped that he would have the horse destroyed for attacking his son. Bonnie answered that you don't let inexperienced kids use spade bits. And you don't drive pitchforks into horses' chests. And that she would take the animal off their hands to spare them problems with the Humane Society.

The man accepted the offer. When she asked for the horse's registration papers, he said that he did not have them. The person who had sold him the horse had never produced papers and, what the hell, it made no difference—the horse couldn't race anymore, could he? If she didn't want the horse without papers, she could forget the whole thing, and he'd call the rendering plant to send someone for the horse.

Bonnie didn't care that much about papers. The horse was unquestionably a Thoroughbred, all wounded 17½ hands of him. When a vet came to treat the injuries, she saw that the horse bore a racer's identifying lip tattoo, which was smudged and illegible. She never learned a thing about his history, but could see that it had been no picnic. He was probably six years old. The fresh wounds inflicted by bit and pitchfork were merely his latest. Near his jugular vein was a perfectly circular scar imprinted, perhaps, by the end of a steel pipe.

And now he was so shocked and exhausted by his fight that

she and the vet were able to trailer him from the battle site to a paddock behind a friendly neighbor's house. For a week Bonnie did nothing but feed him soft mash and leave him alone. He needed a chance to heal and regroup. Perhaps in time she could form a working relationship with him.

He could not be touched. Which meant that he could not be groomed and could be fed only from outside his enclosure. Indeed, he could not be approached. When Bonnie brought his food, he stayed at the other side of the paddock until she left. His experience with the would-be jockey and the spade bit had been the final straw.

Bonnie studied him from a distance. He was an unusually well-made animal. Whoever had sold him to the boy's father had spoken truly when calling the horse sound as a dollar. His numerous bodily scars notwithstanding, he was handsome. He was now a bit down in flesh, of course, but that could be corrected after his tongue and mouth healed.

One morning at the beginning of the second week, Bonnie ambled to the paddock with her portable radio. She sat down and leaned her back against the paddock fence. After an hour or so of quiet music and no talk at all, she unwrapped a piece of watermelon rind, tossed it over her shoulder into the paddock, picked up her radio and left. When she returned an hour later, the rind was gone.

She made a routine of this, staying longer at each visit. Four times a day with radio and rind. No talk. An hour or two of FM violins with her back to the fence. Then a piece of rind tossed into the paddock as she departed. The rind was never there when she came back.

On the fourth day she sat in her usual spot and pulled out the usual piece of watermelon rind and heard the horse walk to the fence behind her. She raised her arm straight over her head, rind in hand. She felt the horse take it.

When she arrived the next morning, the horse was standing at the paddock gate, waiting for her. She put her radio on the ground and stood near the gate, eye to eye with the horse, music playing. She reached into her pocket for a piece of melon. While

the horse watched, she cut away the center meat, doing an entic-
ingly, sloppy, drippy job of it. The horse stretched his neck as far
as possible to bring his nose as close as he could. He wanted the
rind.

She stepped closer, entering what she knew the horse regarded
as his safety space. He stepped quickly backward and snorted in
annoyance at the intrusion. But there she was with the delight of
delights, a juicy piece of watermelon rind. Now she spoke.

"You've got to be the biggest, ugliest moose I've ever seen in
my life," she said in warm, congratulatory tones. "I've been here
long enough for you to know that I'm not going to attack you.
And I know you want this nice piece of watermelon. So come and
get it."

It took him a while, but he actually came and got the rind.
While he was munching, she leaned over and blew gently into his
nose. The formality of introduction. He breathed back.

She continued her four visits a day. His paddock was on a
piece of flat land at the bottom of a 100-foot hill. Every time she
arrived, she found him looking up the hill, waiting. When she
came into view, he raised his head and nickered. She took to
putting her radio down on the hill so that she could bend over,
slap her knees and wave her arms, calling "Hi there, Moose! Hi!"
He liked that. When she got closer and sidled about, arms raised
and head cocking from one side to the other in rhythm, he
would reply horse-fashion, following her lateral motion with his
own prancing feet and rolling head.

About a week had elapsed. She was now welcome in his paddock. They had circumvented the bad memories and had entered
a relationship. She had never touched him.

When she began their work, she treated him as if he were a
foal. She solicited his playfulness, guiding him through the ele-
mentary schooling described in the previous chapter. Whenever
his body language showed that the latest maneuver was not quite
comfortable, she abandoned it for the time being and let him do
something that he enjoyed doing well.

By the end of the summer—six months after his experience
with the spade bit—she and Moose were roaming the hills bare-

back. He had learned all the fundamentals, step by step. He had gone through the L-shaped, pole-lined path forward and backward, through rubber tires, and through rain and mud or under a sprinkler. He responded accurately to all commands when on the lunge line or when doing his thing at a distance of twenty or thirty feet from her, without any line at all. He had accepted the discipline of the ground tie and came to her when called. He had not worn a bridle, bit or saddle except to prove that he would do so when asked. He had felt no whip.

When they were out in the hills, she often let him choose where to go. This led to a game. When he took her under a suitably low branch, she would grab and dangle from it as he loped away. He would screech to a halt, turn around and snort. When she called to him, he would toss his head and return so that she could drop from the limb to his back. The rogue had become a lamb.

She kept him at the Loudon place, where he allowed Hazel to tend him and even learned to accept brief periods with Don when Bonnie was close by. But for the rest of his life, no man ever rode him. And he would charge any male stranger who came too near. If Bonnie was there, she would call, "Moose!" in a low, particular commanding tone, and he would skid to a stop, standing in the posture of apprehensiveness. He never overcame this fear or his aggressive way of dealing with it.

And so he could go to no horse shows and take no trail rides with groups. He and Bonnie roamed the hills alone. He was a cross-country jumper of the highest class. And an all-time great player of games. And a dear friend to her. They were content.

7. BODY LANGUAGE AT
THE RACETRACK

Racegoers who understand equine body language enjoy a tremendous advantage—the stuff of which bettors' dreams are made.

The dreams come true. Competent prerace observers of horses in saddling enclosures, walking rings, post parades and warm-ups seldom mislead themselves into wagers on unready animals. They may attend the races infrequently and, for that reason, lack current knowledge of local horses, trainers and riders. And they may not be especially adept at the intricacies of handicapping. Yet they do nicely in the wagering department. They simply limit their support to unmistakably fit horses that run at good odds.

We have divided this chapter into two sections. The first is for all racing enthusiasts, expert handicappers or not. It equips them with the ability to distinguish the prerace behavior of likely winners from the prerace behavior of likely losers. The second section is for handicappers, or persons interested in that challenging pastime. It probes more deeply, explaining how the prerace behavior and appearance of a horse may illuminate some of the obscure or ambiguous features of its racing record. In other words, it shows how prerace observation reduces handicapping guesswork to a happy minimum.

We begin with a description of the three kinds of prerace body language associated with horses that win at least 90 percent of all races. The information applies with most validity to Thoroughbred and Quarter-horse racing, in which the audience can see the animals in the paddock. Where trotters and pacers are paddocked out of sight of the audience and can be seen only during prerace jogging and scoring (warm-ups), fanciers of harness racing have a somewhat more difficult time.

The Sharp Horse

This is the healthy, ready animal so eager for competition that it may be thoroughly keyed up. It may sweat. It may dance and wheel almost fractiously. Uninformed observers may confuse its body language with that of apprehensive nervousness, but the differences are numerous. They will become clear enough as this chapter proceeds.

Highly keyed and sweating or not, this horse is the embodiment of physical well-being. Its coat shines like burnished furniture and may be noticeably dappled. Its mane and tail gleam. It is neither fleshy nor bony. Its haunch muscles have a particular rippling aspect less frequently seen in other horses.

And again, sweating or not, it is all eagerness. On being led to the paddock past the crowds that line the fences, the animal prances on its toes, head tucked down toward the chest, neck arched, ears pricked forward. Its tail may be held slightly upward in a position of exuberance. It sometimes looks up to the crowd, interested in all the commotion and seeming at times to respond with its own proud strut.

In the saddling stall, the sharp horse is unlikely to be entirely quiet. It is much too full of itself, much too anxious to run. It stands in place just sufficiently to permit the attentions of its handlers, meanwhile raising its head to sniff the air and look for action. Led around the walking ring before being mounted, it positively shows off, dancing on its springy legs, head almost touching its chest. When the jockey mounts, the horse's feet become more restless, the animal recognizing that race time is nearer. The muscles may quiver now, a sign of panic in another horse but not in one that carries itself proudly.

During the post parade in front of the stands, the body language intensifies. The ears alternately turn back to the rider or prick forward. The springy prancing continues and may now include a sideways kind of dancing to face the noisy crowd. The head is down on the chest, neck arched, unless the horse is being escorted by a lead pony and rider. In that circumstance, the observer who uses binoculars may see the eager one pushing its nose against the pony's neck, urging it forward: "Let's get on with it." Sometimes, to spare the pony that kind of hectoring and more often to maintain stout control over the race horse, the escort rider may take a short hold on the lead chain, raising the horse's head and restricting its movements. If the horse's previous deportment was that of sharpness, the sudden elevation of its nose should not be misinterpreted as a sign of distress. Neither should the sight of kidney sweat between the rear legs. This is the sweat of excitement, not fear.

When the lead pony finally goes into a gallop, the sharp race horse usually displays a special and definite kind of body language in its first two strides. The tail goes up, the muscles of hindquarters and legs gather, the hind feet dig emphatically into the ground, and the horse may rear slightly and almost lunge into a canter. If the escort rider permits, the horse's chin will touch its chest under the restraining hold of the rider. The ears continue to alternate between the fully pricked-forward position and the one in which they are turned backward for full attention to the words of the rider. If the escort rider is chatty, the horse's ears may sometimes swivel in that direction.

During the warm-up on the backstretch and turn, the horse is a portrait of controlled strength: neck arched, head down, ears forward, tail up. Waiting to load into the starting gate, it may dance but gives nobody serious trouble. When its turn comes, its ears thrust forward, its nose points straight to the stall, and the animal charges right in, like a hungry horse going for the first big mouthful of a bucket of grain.

Awaiting the start, the sharp horse stands quietly. Its hind feet are firmly planted for forward propulsion when the gate opens. The front feet may be slightly restless. When the bell rings and the stall opens, the sharp horse leaps forward in a long stride.

We have already suggested that an authentically sharp horse need not conform to the foregoing description in every detail and at every stage of the prerace preparations. It may sweat and it may be conspicuously restless. But its body language remains essentially that of sharpness, and its composure tends to increase as the waiting period winds down and the rider brings it closer to the starting gate.

Also, a sharp horse, like any other, can be distracted and upset by external events. For example, some track managements unaccountably place open loudspeakers in their walking rings. When the traditional bugle call "Boots and Saddles" blasts in full amplification at the nervous systems of the horses, all but the dullest of them shy. Should the sharp one's rider yank its head too violently during that flurry, the horse's mind may be distracted from racing to anger. Chances are greater that the animal will be so race-minded to begin with that it will recover from the blare all by itself and that, in any case, the rider will be more help than hindrance.

During the post parade, whether previously blasted by "Boots and Saddles" or not, the horse may be unnerved by a trumpeting band playing immediately adjacent to the point where the animals enter the track. And whether *that* happens or not, the horse may behave fractiously when parading before the stands, half rearing, dancing sideways, even bucking once or twice and swishing its tail in disapproval of continued delays, or din, or whatever. At this point, what differentiates continued sharpness from serious fractiousness is the position of the ears. Those of a sharp horse remain forward in alert interest, or back to the jockey, or turned occasionally toward a jabbering escort rider. Furthermore, as soon as allowed to canter, the animal again becomes the very model of a sharp horse.

But if the ears flatten in real irritation and the tail starts popping up and down, the horse's patience has finally been overtaxed by one or another kind of human blatancy. You had better watch more of the warm-up to see whether it recovers its equanimity or blows up, gets severely overheated and becomes an unattractive risk.

Later, a sharp horse may be undone when manhandled for no good reason by an assistant starter, although perfectly willing to enter the gate unaided. One of the most flagrantly destructive situations of this kind occurs at the few tracks that permit an official to stand behind the gate, promiscuously flicking a ringmaster's whip at the heels and hindquarters of horses. If that chucklehead decides to whip a sharp horse as it charges into the gate, the sharpness may vanish at that instant.

An interesting sidelight is the behavior of the pony that escorts a sharp horse. Lead ponies prefer nice, quiet horses that give them no trouble. They generally regard sharp horses as overeager nuisances. They do not like to be prodded and pulled. You may well see the sharp horse's pony glaring angrily, muzzle curled, ears flattened, tail popping.

Of all races in which a sharp horse can be taken at face value and wagered upon, maiden races seem to provide the most glittering opportunities. The sharp specimen usually is the only such animal in a generally demoralized group of apprehensive non-winners. In any case the racegoer who wagers on none but sharp horses has a splendid chance of ending each season in the black—even if the body language of horses is the sole basis of the selections. However, to eliminate many sharp horses that lose their races, it is necessary to become a handicapper, combining the principles of that exhilarating hobby with knowledge of body language. Of that, much more later.

The Ready Horse

What distinguishes this potential winner from the sharp horse is a lack of extreme eagerness. Where the sharp horse is healthy and impatient, the ready horse is healthy and content. Its coat may gleam but seldom has the matchless bloom so often associated with sharpness. The behavior is quieter, fully tractable and less hectically obsessed with getting the race under way. The horse glances at the crowd, but its attention does not linger there, returning back just as quickly to the handler. It dances less than the sharp horse and sweats less if at all.

The ready horse stands quietly while being saddled. If the handler at its head is scratching its ear or jaw, the animal may actually drop the head and enjoy the attention, its mind entirely off racing. When led back around the walking ring to be mounted, the horse walks out willingly, with little dancing or tail movement. Its head will be foreward or sideways, its ears flicking in response to noise here or activity there.

The horse joins the lead pony willingly, moving nicely, neck not highly arched, chin not touching chest. Its rider's own body language is a lot more relaxed than when mounted on a sharp horse. The rider holds the reins more loosely, sitting comfortably in the saddle, not tensed to restrain the horse. You will probably notice a good deal more talking and joking between jockey and escort rider than is possible when a sharply eager horse is involved.

You will also notice that the escort rider has a relaxed hold, and the pony's body language is that of calm acceptance, head down, ears flopping lead-pony style, as if delivering the milk.

When they go into the prerace canter, the lead pony may actually stride out before the race horse does. The racer may trot for the first fifteen feet and then move smoothly into a slow, collected canter with head down. The jockey will not have to stand up and lean back on the reins to maintain control. This horse can be managed from a sitting position.

The ready horse enters the starting gate willingly, but does not lunge into it like the sharp horse. Once in the stall, it stands fairly quietly, perhaps shuffling its feet somewhat. Its hind feet are seldom firmly planted.

This is the kind of animal that wins most races, simply because truly sharp horses are not as numerous. Although the behavior and appearance of the ready horse seem unimpressive by comparison with those of the sharp horse, they are far more reassuring than those of the other kinds of animals we shall now discuss. Furthermore, occasions do arise in which the merely ready horse

defeats the sharp one, and deserves to. Such an outcome may be
one of the vagaries of racing luck but more often is not. Readers
who learn how to handicap will often be able to tell when a ready
horse of intrinsically superior quality is likely to defeat a sharp
horse. Health and eagerness are not always enough.

The Dull Horse

This may be a perfectly respectable animal not yet in top form
after a lengthy absence from competition, or beginning to lose
its form after too much racing, or suffering from the effects of a
minor ailment like a low blood count or the beginning of a mild
infection. Its coat seldom gleams and may even look slightly
rough. It moves willingly enough but in a flat-footed walk without
springy movement or high leg action. The neck is not arched.
When the horse walks past the paddock crowd, its ears may flick
toward the hubbub, but the animal does not necessarily turn its
head in interest. Any display of interest is brief. At this stage you
may notice the light beginning of sweat along the neck, but no
such sign in the kidney area between the hind legs.

During the saddling, the horse's head remains down in a natural, relaxed position. The ears are sideways, pointing downward.
The handler who might be standing at the head of a livelier
animal probably does not even touch a rein in this case. If a
neighboring horse kicks a stall wall or raises some other kind of
rumpus, the dull one will raise its head and turn its ears toward
the sound, but that's all. Indeed, this horse may seem at times to
be dozing in the saddle stall, its head down and weight shifting
from one hip to the other.

Unlike a sharp or a ready horse, which usually shies at an
overamplified "Boots and Saddles" or other sudden blare, the
dull one may do no more than swivel its ears toward the disturbance. When conducted to the lead pony, it may greet that animal
with a low nicker or simply nose its neck. During the ensuing post
parade, the jockey may notice that the horse is moving either at
a flat walk or in an actual shuffle and may try to rouse it by
rattling the bit, shifting weight in the saddle, slapping it on the

neck barehanded, standing up, sitting down, yelling and even using the whip.

The escort rider has a slack hold on the lead rope or chain, which swings loosely. The pony is probably ahead of the race horse. When warm-up time arrives, the pony starts cantering and the horse only musters an awkward trot. It canters when tugged by the escort rider and kicked by the jockey. Its head remains down but the neck is not arched. Ears generally are turned toward the jock. At any opportunity to stop and stand absolutely still, the horse does so.

This is the kind of horse you see taking short speed dashes on the backstretch, having been busted with the whip. But the lead pony is quite likely to have its head in front of the race horse's even then! When they reach the starting gate, the dull horse probably hesitates just long enough to earn the physical attention of an assistant starter. In the stall the animal stands flat-footed and can be relied on to be among the last to leave the gate at the start of the race.

He is not sharp or ready. He is a deadhead. But if he is in a field of equal or lesser racing ability without any sharp or ready opponents, this dull horse may win. His chances improve if his opponents include only one or two other dull ones and all the rest are angry, frightened or hurting.

So much for the kinds of horses that win nine races of every ten. Now let us consider the kinds of horses that win only by accident.

The Frightened Horse

The animal may be frightened before arriving at the paddock and may remain frightened all the way to the starting gate. Or it may become frightened at any time. When fear persists, no reasonable person should expect the horse to win. Fear consumes energy needed for the race. Furthermore, the mind of the frightened horse is not on racing but on survival. The animal's fear may be specifically based on unpleasant memories of previous experiences in competition. Equally often, a horse suffers from the apprehensiveness of inexperience. Fear is also seen in horses returning to action after lengthy absences.

The body language of a frightened horse is unmistakable. The animal comes to the paddock with head high and in rapid motion, eyes rolling so violently that the whites are visible, nostrils flaring, ears constantly flicking in all directions and often unsynchronized with each other. It may whinny or neigh. Its jaws may work rapidly, the mouth opening and shutting. Its leg action is high and erratic. Its tail swishes from side to side and up and down, but it does not pop like that of an angry horse. This animal is too panicked to organize anything as emphatic as a pop. It has already begun to break out in sweat at the neck and shoulders, between both pairs of legs and at the flanks.

The handler has the horse under tight hold, usually with a stud chain over its nose, under its lip or straight through its mouth. The horse typically moves in a half circle, crossing completely in front of the walking handler, fighting the chain, reaching and flailing as far as it can with its legs, trying to flee.

It reacts strongly to any noise, any sudden movement on the periphery of its vision. Each new stimulus increases its fear, as do the routines of prerace preparation. The animal shies from the crowd, repeatedly entangling the handler. It may break into a desperate trot to get away from the onlookers, then wheel or pivot to face the danger, retreating backward.

The terrified animal probably enters the saddling stall eagerly, as if into a sanctuary, spinning immediately to face the noises. It cannot stand still in the stall. It moves all over the place, probably backing into the rear wall, then coming forward and kicking it. The volume of sweat increases until the horse is drenched. Its eyes are still rolling, ears flicking, nostrils flaring. And its teeth begin to chatter.

If all this is not sufficient to mark the horse as seriously distressed, the handlers now reveal the truth in their own ways. Extra people materialize in the stall to help the trainer and groom with the problem. This is the kind of horse frequently seen being led out of the stall, circled and led back again in an attempt to use up some of the restlessness. It seldom works. When the handlers try to fasten the saddle, the horse resists. It also resists being led to the paddock for mounting. The sweat continues to drip. Waiting to be mounted, the poor horse may stand quaking, with all four legs spraddled. When the jockey's weight lands on its back, it half rears, wheels and circles as the rider tries to get hold of the reins. A horse in this condition may actually fall down in the walking ring, especially when its fright is compounded by a sudden noise or by accidental collision with another horse.

On the way to the track, the jockey has the tightest possible hold. The lead pony may have a somewhat calming effect, but it is too late. The horse's energy is depleted, its race lost. During the parade, it likes to touch as much of the pony as possible, its head over the pony's neck for solace. If the escort rider does not like that and prevents it with the lead chain, the frightened horse's nose is high, eyes and ears still in motion, legs flailing every way but straight ahead. The front legs move with a particularly high and uncoordinated action. The hind legs tend to be well underneath the body, supporting most of the weight while trying to jump sideways. The lead pony tries to proceed straight down the track, but the race horse moves in small, spastic jumps at an angle to the pony.

When they reach the backstretch, away from the crowd noise, the horse may settle a bit, start to drop its head and move fairly straight. But the proximity of the starting gate always starts a fight. No competent starter ever permits assistants to whip a frightened horse, but incompetent ones do, and this always makes things worse. In the end the horse is practically lifted into the gate.

This is the kind of animal most often involved in starting-gate disasters. These are the ones that rear and throw their riders, or even go down themselves, or hang themselves over the partitions between gate stalls. Sheer accident determines what happens to them at the start of the race. If they happen to have their feet well placed when the gate opens, they come out first—fleeing, ears laid back, nose out, jockey hanging on for dear life. They seldom hold their speed into the homestretch, invariably finishing in the rear of the field.

Those that do not happen to have their feet under them at the start may break last from the gate and try to run directly toward the inner or outer fence. After being straightened by the riders, they show tremendous bursts of speed which enable them to catch up with the others but leave them thoroughly exhausted and out of contention.

The reader should bear in mind that any horse may become frightened at any stage of the proceedings. However, the more ready to race, the more easily a horse withstands the effects of a frightening sight, sound or event. Some horses do not become frightened until saddled or mounted, as if they did not believe that anyone would put them in this desperate situation. They presently become as depleted as those whose fear began earlier in the day. Some horses that seem quite composed in the walking ring and post parade go to pieces at the sight of the starting gate.

Whenever it occurs, the onset of the body language of fear should alert the observer to the existence of a severe problem. If the horse does not recover promptly, its mind is not on racing, and the smart bettor will be guided accordingly. Unfortunately, fear may set in after the race begins, when it is too late to avoid betting on the animal. A previously sharp-looking horse may get the early lead, stumble and quit cold, having been terrified by the possibility of falling. The same may happen to a horse intimidated by racing in close quarters or being bumped sideways.

The Angry Horse

Under this heading we now consider horses of assorted ill temper, from the mildest irritation to the wildest fury. The most familiar type is the sour animal easily provoked by prerace occurrences. This individual may dislike anything that reminds it of racing, as Silver Series did for a while. Or it may feel a special grievance against a particular handler, rider, assistant starter, kind of weather or racing surface, or any other aspect of the day's business, including some turn of events during the race itself.

The body language and underlying problems of the angry horse differ from those of fear, but the end result is identical.

Having expended precious energy before the run to the finish line, the horse loses. The only kind of angry horse with a reasonable chance to win is one whose prerace mood never exceeds irritation, who settles into a more cooperative frame of mind long before post time and makes little or no fuss at the starting gate.

By contrast with a frightened horse, the angry one rarely sweats. In moments of annoyance, its ears flatten back against its head. Otherwise, they tend to be held straight forward. Indeed, the degree of anger is indicated by the share of time the ears spend in the flattened position, and by the distance between them and the head. Mildly flattened ears bespeak a lesser anger than when they are pinned so tensely that you can scarcely see them. When its ears flatten, the horse simultaneously raises its tail and pops it down against its hindquarters, or slaps it at whistling speed from side to side.

Whereas the eyes of the frightened horse may roll, displaying white, those of the angry horse take on a fixed glare. The upper lip may curl, much like that of a dog about to bare its teeth in a snarl. And the legs and feet of the angry horse show none of the uncoordinated flailing associated with fear. Instead, the animal's movements are quite deliberate. The carriage of the head is also different. Instead of being in the panicky, nose-upward position, it spends energy in irritated tossing, trying to wrench the lead rope from the handler or the reins from the rider.

When a racer moves by the crowd for the first time, en route to the saddling area, the observer can tell whether (a) it is irritable or not and (b) whether its problem relates to a dislike of crowds or noise or both. A horse that virtually ignores the crowd, responding to the noise with nothing more than a flick of the ears in that direction, certainly is not a sharp animal and may not be an angry one either. But if it stops, tries to face the crowd, flattens its ears, pops its tail and even kicks, you know at once that it is no fan of the fans. This alerts you to watch the prerace formalities closely and see whether the annoyed one gets over the annoyance, or just gets angrier.

In the saddling stall, an angry horse whose ire relates to provocations other than saddling will merely look irritable, dancing restlessly, ears flattening briefly, feet occasionally kicking the rear wall. It may also curl its lip and snap its teeth. But if saddling is one of the causes of the horse's negative temper, it pitches a battle. This is the horse that you see dragging human beings around outside the stall as they try to fasten the saddle. You may even see one of them twisting one of the horse's ears—or even biting it.

When the angry horse goes to the walking ring to join its jockey, it moves not at an ordinary walk but at a tensely quick trot in which the hooves hit the ground hard enough to leave unusually deep prints. The animal is literally taking out its resentment on the ground. Its tail hisses through the air on both sides of its rump, or pops vertically, or both. Its head activity also may become quite elaborate as it tries not only to rid itself of the lead rope but to snake its head around and bite the handler. And you may also notice a considerable amount of kicking toward the crowd or perhaps toward another horse too close behind.

Nearby horses, who might have been perfectly composed in their own saddling stalls, may now show the body language of fear, eager to give the angry one a wide berth. And the angry horse's handlers present the body language of human anxiety. They are notably light on their feet, paying strict attention to the horse's every move.

An impatient handler who now punishes the irate horse by hitting it is almost sure to arouse additional anger in the animal. When this flurry immediately precedes the ascent of the jockey, as it often does, you can expect the horse to blow its cork, rear and do its best to throw the jockey.

In the post parade the animal's annoyance may center on the lead pony, the race horse immediately behind or the fact that it is not allowed to march at the head of the procession. It may try to bite the pony or the escort rider. It may kick and pop its tail. Its ears surely will spend much time in the flattened or fully pinned position. Horses of this disposition are often excused from the parades so that their riders can try to burn off some of the anger in exertion. When next observed through the binoculars on the backstretch, they are often sweating heavily from the physical exertion, not from their anger. If they still display signs of irritation, they are bad risks at the betting windows. In fact, if their previous anger has been both emphatic and prolonged, they may already have expended too much energy. As a useful rule of thumb, which works more often than it fails, you might consider not wagering on any horses whose anger persists through two or more of the following prerace phases: The walk to the paddock; the saddling; the walk to the jockey; the mounting; the post parade; the warm-up; the starting gate.

After displaying its angry fractiousness at the gate (watch the tail and ears and the attention paid it by assistant starters), the horse races much as a frightened one does. It either breaks promptly and shows brief early speed before tiring, or it leaves the gate slowly, expends what is left of its fuel in an effort to catch up and then tires.

One final note. If the horse was merely irritable before being mounted but spends the remainder of the prerace period trying to get rid of the jockey, you can assume that it either dislikes all jockeys or has a particular antipathy toward the present one. You can conclude with high certainty that it will not perform well.

The Overheated Horse

Some horses can ship from the wintry North to the torrid South and promptly run their best races without a period of adjustment to the heat. Most cannot. And, North or South, most horses do not race well when the Fahrenheit exceeds 85°. In really hot weather, with temperatures at 90° F or higher, it is commonplace to see not a single sharp or ready horse during the full course of a racing program. They all look dull and overheated. If the heat is complicated by high humidity, they look even less impressive.

The overheated horse moves as slowly as possible. Its chief characteristic is lethargy. It washes out in sweat, head down, flanks moving rapidly, nostrils flared. But it displays no other characteristic of fright unless it happens also to be frightened, which would make its prospects even poorer.

On very hot and humid days, one also sees an occasional horse kicking a hind leg but displaying no other sign of anger. On closer observation, notice that it appears to wind up the hoof with a horizontally circular motion before kicking out at a diagonal. Its problem is dripping sweat, which tickles the insides of its thighs. It kicks to shake off the discomfort. Its ears do not lie back. Its tail does not slap angrily against its rear.

The Cold Horse

Asked to race at a Northern track in winter too soon after arriving from the warm South, the typical horse is too busy freezing to compete successfully. Its teeth chatter, its eyes are half closed, its ears flopped, its head down, hunkering up against the cold, much as a human would in comparable circumstances. Its only sign of relationship to the upcoming race will be the annoyed slap of its tail when the jockey mounts. This horse is a loser. Even if a racer has never known a warm climate, it is a certifiable loser whenever it finds a given day too chilly and demonstrates that feeling in its body language. Look for a horse that seems less distressed. Perhaps one warms itself during prerace exercise and looks comfortable. If you cannot find such a horse, why bet?

The Hurting Horse

The body language of pain (which we described in Chapter Two) is rarely seen during racing programs and, even then, is expressed too subtly for any but the most experienced observer. The reason is that horses in conspicuous pain are not often required to race, except in such places as permit the use of anti-inflammatory or analgesic drugs which suppress the language of pain along with pain itself.

Having said that, we recommend that the reader suspect the worst of any horse that appears not just dull but ultradull. Its eye is dull. Its coat is dull. Its motions are unusually subdued. It moves in a slow, collected way. Its head tends to hang and its ears droop. Its tail moves but slightly if at all. Its feet often kick up dirt as it shuffles through the post parade. When the prerace cantering starts, the escort pony leads the way. The race horse goes into a stiff trot, with head dropping abnormally low—perhaps as low as the knees. Finally cantering, the horse's stride is awkward, stilted and choppy, with a lot of head-bobbing.

The horse may warm out of its discomfort before reaching the gate, and will show that it feels better by raising its head and striding smoothly. If not, it may become angry and try to stay out of the gate. During the race, if it is still in discomfort, it will move with a low, bobbing head and will lose.

Another kind of distress which might as well come under this heading is that of the horse abnormally light in flesh. The characteristic sunken flanks, prominent ribs and hip bones, and tucked-up abdomen may be due to dehydration, undernourishment, overwork, nagging pain or illness. If distressed body language accompanies these other signs of impoverishment, the horse is not likely to run well. On the other hand, some extremely skinny horses win races. The frequent racegoer learns who they are. The more casual racegoer should probably refuse to wager on any underweight horse that displays any body language other than that of readiness to run.

Rain, Slop, Mud

Most horses enjoy light rain. But few like a downpour. When pastured, they collect themselves in a nose-to-nose circle under the cover of trees and wait it out. Only an eccentric horse disdains the shelter and frolics all by itself in heavy rain.

On a very rainy day at the track, most of the horses would rather be nose-to-nose under trees. They slouch around the walking ring and through the post parade with eyes half shut, ears down and back to keep out water and wind. They look toward the barns and nicker.

They particularly hate wet tails, which become uncomfortably heavy and, when swished for normal purposes, can sting the horse's own flesh. Even when a race horse's trainer wraps its tail to keep it dry in rain, the animal instinctively tenses its haunches and upper thighs, clamping the base of the tail close to the hindquarters, and tries to tuck the member between its legs for shelter.

If you notice a horse behaving contentedly in heavy rain at a track, it almost surely is the one to back. Watch the prerace canters. Most of the horses will be trying to hide in their lead ponies' manes. But yours will move right out, as if in its natural element. Instead of the body language of discomfort, this is the body language of playfulness.

Rain or not, body language also helps you to pick the right horses on days when the racing surface is sloppy or muddy or, on turf courses, soft. The horse may look dull in the saddling area, but if it moves comfortably and naturally through mud, it becomes a far better bet than the sharp or ready ones that go to pieces when they find themselves on an abnormally soft or slippery surface. The uncomfortable ones walk daintily, as if on eggshells, or like small children picking their way through a wormy garden. They lift their feet quickly, dancing sideways, tails popping, ears back or flicking, heads up, eyes rolling.

But a real off-track runner concentrates on the task at hand, head down, ears forward. The veteran observer will notice that this animal's stride is slightly more deliberate than normal. It plants its hind feet squarely for extra stability and moves through the muck with a balanced, determined stride. The horse wants to run on this footing and gives every sign of ability to do so.

FOR HANDICAPPERS ONLY

Let us reconstruct a familiar situation. You have handicapped the race. It seems to you that three of the horses have approximately equal chances. Exercising your knowledge of equine body language during the prerace formalities, you find that each of the three is authentically ready to run. No other horse in the field looks as good, either on paper or in the flesh.

Where does this leave you? It all depends on your individual style. Perhaps the record of one of the animals indicates recent improvement, suggesting that additional improvement might make it a winner today. Or perhaps the contenders seem so closely matched that you either pass the race or bet on the one whose odds are most attractive. All right, this race is not an easy one. In others you will be able to separate contenders on the basis of their body language. Opportunities of that kind will be numerous enough to make each of your racing excursions more exciting and more enjoyable than ever before. What could be more exciting or enjoyable for a handicapper than the ability to see things that are invisible to others?

Records—Published and Private

While you agonize over the race in which your handicapping finds three evenly matched contenders and your prerace observations produce no reason to prefer one of them, other members of the audience may be burdened by no such dilemma. For example, Handicapper "A" may be at the races every day and may mark the program with his evaluation of each horse's prerace demeanor. Naturally, he saves those marked programs for future reference. He probably prefers one of the three leading contenders in today's race because he knows that it seems fitter and perkier today than on other recent occasions. Since the horse's previous performances were every bit as good as those of the other two contenders, today its improved mental and physical condition should be a winning advantage.

The marked program is an enormous help. But not everyone can get to the races often enough to compile records of that kind. The remainder of this chapter suggests means of using published records to greatest advantage in an approach that combines handicapping and knowledge of horse language.

North American handicappers get their information from the past-performance records, race-result charts, workout tabulations and statistical summaries published in *Daily Racing Form*. Although details vary from place to place, handicapping materials comparable to those of the *Form* are available in all major racing centers of the world. Before launching this discussion of ways in which formal handicapping combines with prerace observation of horses, we must point out that the track itself is seldom the right setting for a study of horses' racing records. That part of the selection process should be attended to before one arrives at the track. A handicapper who hopes to pay informed attention to horses in the paddock, post parade and prerace exercises will be unable to do so unless the horse's records have been analyzed and interpreted beforehand.

The Effects of the Latest Race

Few winning horses materialize out of the blue. Most show signs of improvement on the track before actually winning. In studying the records, the handicapper's highest priorities include the search for horses whose form seemed to improve in their most recent starts.

Did the horse improve even slightly? Was that race its first, second or third after a protracted absence from competition? It ought to improve again today. If it is entered at a suitable distance on a comfortable surface with an adequate rider and is at no disadvantage in running style, weights or post position, it might be a winner. You therefore will be eager to see the animal before the race. If it looks sharp or ready, you may have something.

The same will be true of a horse whose latest race was a predictable victory or an impressively close defeat in which it showed a good deal of early speed or a powerful finishing run. In short, the horse has already established its good form and need only remain in form of that kind to be a threat today. If it turns up with the demeanor of a sharp or ready horse, you know that it is a serious contender. On the other hand, if it seems dull or in distress of some kind, that last effort or a more recent training occurrence may have knocked it off form.

Our friend who is there every day, marking every program with analyses of the horses' behavior before their races, can tell you whether this horse is the unusual kind that runs well even after looking dull or distressed. If you cannot find him or if he refuses to talk, you can console yourself with the knowledge that no law requires you to bet on every race. If this one is too puzzling, abstain. Before deciding to do so, however, you might look around the field to see whether any of the starters looks downright sharp, has a respectable record, seems to be properly placed as to distance, class, etc., and is going at nice odds.

Consider now the horse whose latest race was a poor performance untypical of its previous history. For instance, the horse was the betting favorite in the race but was thoroughly beaten, finishing in the second half of the field. In each of its previous six races during the current season, it had finished first, second or third. No explanation for the bad performance is found in any accounts of the race. Perhaps the footnote comment in the *Daily Racing Form* result chart says that the animal "never threatened in a dull effort." But why?

A competent handicapper routinely checks to see whether a horse's defeat can be explained away (and forgiven) on grounds that the animal was entered at the wrong distance, or on the wrong footing, or in the wrong class, or with the wrong jockey, or under oppressive weight. Or whether the horse had been away from the races for months and should not have been the betting favorite to begin with. Or perhaps the horse was a new arrival at the particular track and had not yet adjusted to the surroundings.

Finding no such excuse in the record, many handicappers give the horse the benefit of the doubt. They reason that horses are not machines. All living creatures are entitled to an occasional bad day. And so forth. The reasoning is plausible but entirely too easy. A handicapper unable to find an excuse for a horse's otherwise inexplicable defeat must always respect the possibility that the loss may have been thoroughly inexcusable and that the horse may run even more poorly today.

After all, no race horse's good form is permanent. The inevitable loss of form derives most often from the exertions of racing and training. Some horses, especially younger ones, hold a sharp edge through five or more hard, driving finishes. Others, especially older ones, come unraveled after two or three stern efforts in close succession. If the horse's record is one of recent and repeated all-out effort, and if the animal appears in the paddock with the drawn, tucked-up look of overwork, it probably will lose. We concede that this individual horse may be the exception that wins even when it looks awful. But that kind is severely outnumbered by those that look like losers before losing and like winners before winning. Putting it another way, if the horse has been exerting itself more often than some horses can tolerate and if it looks tuckered out, expect it to lose.

Back now to the dilemma in which the record offers no reason, excusable or not, for the defeat of a horse that had been winning or running close to winners in its recent performances. If it looks like a wreck in its prerace exercise, you will undoubtedly accept that evidence at face value and be correct more often than wrong. But if it looks ready to run, the dilemma thickens.

If you assume, as most bettors will, that the animal simply tossed in a bad race and will be back to normal today, you are guessing and will be wrong more often than right. There always is a reason for equine behavior. If you assume that the horse's recent defeat was a result of an intimidating occurrence that escaped the attention of those who reported the running of the race, you have an excellent chance of being right. At some point the horse was bumped, crowded or kicked. Or it slipped or stepped in a hole. The younger and less seasoned it is, the less probably it will recover promptly from the frightening experience and win its next race. Even in the unlikely event that it looks sharp or ready before entering the gate, it may go to pieces during the race, when it arrives at that part of the track where the unfortunate happening took place.

If the published record of a horse's latest race includes explicit information about intimidating difficulties it may have encountered, the handicapper is spared the strain of guessing. In the above paragraph, little justification was found for assuming that a lightly raced young horse could snap back to form after trauma of that kind. In the present case, with the untypically poor performance explained in print, even less justification exists. The handicapper should expect the inexperienced two- or three-year-old to perform below standard for two or three races before recovering its aplomb. Those who take that long far outnumber those who shake off the ill effects more quickly. As to older horses, a really tough one who looks sharp or ready can be given the benefit of the doubt—especially if the odds are generous enough to repay money lost when such a horse remembers the upsetting bump and turns timid during its next race.

For novice two- or three-year-olds, one of the most intimidating experiences is completing the turn for home and literally colliding with the overwhelming phenomenon that jockeys describe as "the wall of noise." The problem is not just the volume of sound that issues from a crowded grandstand at that stage of a race. Even more frightening is the emotional intensity of the vibrations generated by thousands of human beings in a state of high excitement. If the young horse is the least upset before its race (as so many are), the "wall of vibes" is almost sure to complete the job. The youngster may enter the final straightaway in the lead, but probably chucks it at once, its mind distracted from the job. This is as firm a reason as any for vowing never to bet on an apprehensive two- or three-year-old. It also alerts the handicapper to study the records of such horses as they acquire seasoning. After perhaps three or four unsuccessful starts, and having begun to develop a reputation for faint-heartedness, the fledgling carries its speed a bit farther into the stretch one afternoon and when it comes to the paddock for the race after that, seems calm if not actually sharp. It has overcome its difficulties with the crowd vibrations, which it now accepts along with the other incomprehensible discomforts of its trade. If it is the only ready horse in the field, it almost certainly will win.

One of the most subtle and least noticed problems in the sport is the deliberate intimidation of one horse by another during a race. Some highly competitive animals are so reluctant to let another pass that they swing their heads over and bite. This is known as savaging, which is against the rules, can lead to disqualification and is prevented by jockeys able to anticipate it. Unfortunately, a dominant horse can intimidate another without actually biting. Flattened ears, slightly swinging head, glaring eyes and popping tail are components of a message that never fails to impress a nearby horse. If that horse feels subservient, it will back up—even though it might actually be able to win the race if unintimidated.

Handicappers are advised to watch the videotape reruns of races displayed on monitors at many tracks. See whether the winner flattened the ears and popped the tail just instants before a neighboring runner tossed in the towel. If so, do not expect the intimidated horse to offer serious opposition when the two race against each other again. And note also that persons who complain that the loser's jockey "did not persevere" in the stretch would do well to consider that whipping and other "perseverance" are of no more avail on an intimidated horse than on a horse with a broken leg.

Of all negative experiences, the worst is a stumble or an actual fall. The next worst is the sudden need to avoid trampling on a fallen horse or rider. No horse can reasonably be expected to perform well in its first race after an occurrence of that kind. As we emphasized at the start of this book, horses perceive loss of balance as a threat to survival.

A young filly brought back to the races within a week after stumbling or falling on the track will probably resist efforts to push her into the starting gate and probably will need at least three starts before recovering a willingness to race. Even if she is spared the ordeal of competition for weeks after her near-disaster, she will show reluctance in her next outing or two. For that matter, so will older fillies and mares and males of all ages. Which is why it is so important to read all race-result reports carefully, noting the names of horses that stumbled or fell. Expect nothing from them in their next races. And expect nothing from the poor horse that was running directly behind a horse that fell. If the chart shows that your horse was racing forward of the accident, the animal will probably be unimpaired.

When a horse unseats its rider at the start of a race and tries to run with the rest of the field, it may traumatize other horses with its hectic swerving. The reader will profit from an effort to ascertain whether the loose horse actually bothered any of the others. When one of the bothered ones turns up for its next race in a nervous sweat, you will understand why.

The Effect of Weight

The horse's record shows that it has run its best races when carrying 116 pounds or fewer. Whenever saddled with 120 pounds or more, it has finished out of the money. Today it runs at its favorite distance against horses of a class that it can usually handle. But it has been assigned 121 pounds. Watch closely for signs of irritation in the post parade. We shall not trap ourselves into argument about whether a horse can tell when it is carrying five more pounds than it likes. But we can state with confidence that the extra weight often reminds the animal of previous occasions in which it was driven to the outer boundaries of survival when carrying a burden that felt like today's. Moreover, if the rider is light and the saddle pad is laden with weights to bring the impost to the required 121 pounds, the horse may begin fussing as soon as the heavier saddle is fastened to its back. This effect—and its cause—are unmistakable when the horse arrives at the saddling enclosure cool, calm and collected and does not rebel until it feels the extra weight. Needless to say, this behavior substantiates the record: The horse cannot win under the weight.

The Jockey Angle

The horse's record establishes it as a strong contender in its race. It looks and acts the part while being led to the paddock and while being saddled. But when the rider mounts, the horse becomes upset. The problem is not one of high assigned weight. It therefore may relate to the horse's dislike for the individual jockey. What is their history together? Does the horse usually run well for the rider? Is the rider a busy whipper? Have the whipper and horse been in a couple of all-out stretch drives recently, and does the rider's presence remind the animal that it is about to suffer pain? The reader should be dubious of the chances of any horse whose behavior takes a turn for the worse as soon as the jockey comes aboard. Indeed, the reader should expect this to happen if the horse's record shows that it wins only for one particular rider, yet today will be steered by a jockey for whom it always loses. The horse's attitude is a virtual guarantee that it will continue to lose under this rider. And if irritation or panic persists into the post parade and warm-ups when the horse is mounted by some rider for whom it has won only under severe punishment, it is probably a loser today.

The Long Layoff

Teachers of handicapping emphasize that lack of recent races is an unpromising sign. Whatever else may be wrong with a horse that has raced in the past ten days or two weeks, it has been in competition and should be in tighter physical shape than a horse that has not raced for months. The principle works quite well. All studies indicate that horses absent from the races thirty days or less win somewhat more than their share of races, whereas horses away a month or more win less than their share.

On the other hand, some stables seem to win with horses after layoffs of months. Handicappers learn which trainers manage such feats. The reader of this book can do the same but can go farther. If an animal has not raced for months, it may turn up with a large belly and an apprehensive attitude. Until it races off some of the excess weight and becomes mentally adjusted, it will lose. But if a horse that has been on an extended vacation looks sharp or ready and, in addition, comes from a barn that sometimes wins with absentees, all systems are "Go."

The Battle of the Sexes

Perhaps because they are entrusted with the perpetuation of the species, female horses seldom race with the reckless impetuosity of males. The blinding burst of speed is a specialty of colts and geldings. Fillies and mares seem to be governed by a more keenly developed sense of self-preservation. They tend to spare themselves.

Why, then, have Dahlia, Allez France, Waya and so many other great fillies and mares defeated the best males of their generations in races at classic distances on grass? The answer is simply that such races favor the running styles and temperaments of females. Some of them can run all day, galloping along at a characteristically even pace and retaining ample energy for the crucial final stages. Not many males are so patient. The same sort of thing happens in trotting or pacing races. Horses that pull their drivers and sulkies at a steady pace enjoy a distinct advantage in harness racing. Which is why females do handsomely in that sport.

The kind of running race in which females are at the most disadvantage is sprint and middle-distance competition on dirt tracks—the racing that predominates in North America. This is only partly because the burst of speed is a powerful asset in racing of that kind. After all, some females are capable of high speed. And an even pace wins many a sprint and middle-distance race. The main reason why fillies and mares rarely compete against males on major North American tracks is that the economic realities of the sport discourage such confrontation. The breeding industry uses at least fifty times as many mares as stallions. This means that the market value of a racing filly or mare reflects not only her potentiality as a purse-earner but her prospects as a broodmare. For all colts and geldings below the top level, however, market value depends entirely on racing ability. Thus, a $20,000 colt has more pure racing ability than a $20,000 filly. And since eligibility for entry in most races is stated either in terms of real or fancied market value (as in claiming or selling races) or purse-earnings (allowance races), the females have almost no opportunity to race against males of equal or lesser ability.

A $20,000 filly probably can defeat a field of $10,000 colts and geldings with great ease, but her owner will not risk losing her on a $10,000 claim. And a filly somewhat too good to run in claiming races for her own sex is not quite good enough to compete against the males that enter comparable non-claiming races for their own sex. At the very top of the hierarchy, where North American championships are settled at middle distances on dirt, the male champions usually are better at that game than the female ones.

In many parts of the world, and now we get to the point of this section, females and males continue to race against each other. The practice is standard at minor tracks in North America, where fillies and mares win dirt-track races at all distances when competing against males whose low market value does not exceed their own. At these levels of competition, none of the horses is a breeding prospect. The $2,000 filly is just about as good a runner as the $2,000 gelding. When the leading contenders in a race include members of both sexes, the handicapper should base the final decision on observation of paddock and post parade, where the battle of the sexes may well affect the outcome of a race.

Is one of the starters a hard-hitting old race mare who has already defeated males? Is one of her prime competitors a three- or four-year-old colt? If the colt comes out strutting like the boss of the world and annoys the old girl in the walking ring or post parade, and if she flattens her ears, swings her head and pops her tail at his approach, his strutting will diminish at once. She probably has him beaten. The same thing may happen during the actual race. If so, she will intimidate him in subsequent races if she is able to get near him. His only hope is to leave the gate in a hurry and run away from her. She has awakened his memories of female authority.

On the other hand, if a colt encounters a romantically inclined filly in the paddock and you see them extending their noses toward each other, you can write them off. Their minds are not on the race.

And now for an example of equine sexuality seen at all tracks. A three-year-old colt arrives from the barn to be saddled. He shows all the signs of incipient studdishness—the thickening neck and shoulders, the proud prancing and nickering to announce that here comes the hotshot of the group. A couple of horses later comes another colt, like-minded. The newcomer nickers, "I'm the hot stuff here. Anybody care to argue?" The first one bellows, "You better believe I do!" Or noises to that effect.

Now the uproar. Both colts kick their stalls and give their handlers a difficult time. The rest of the field becomes apprehensive, wanting to stay clear of the trouble that surely will arise if the two colts get close to each other. If the rest of the horses are three-year-old geldings and fillies, the race virtually ends before it starts. The two studs race head and head on the lead from beginning to end, with the others at a respectful distance. But if the field includes an older male or mare, the probabilities are that neither will be intimidated by adolescent silliness. If either is a closing type of runner and in decent form, it will nail the two hotshots in the stretch, beating them to the finish line.

The Generation Gap

An inexperienced three-year-old gravitates to an older animal for leadership, protection and comfort. If an elder happens to be entered in the same race and answers the youngster's apprehensive whinny with a nicker as they parade around the walking ring, the nicker means, in effect, "It's all right. I'm here." If the older one can run at all, it will defeat the younger one. You can bet on it.

Things of that kind occur daily and represent a biological fact which is not nullified by the regularity with which really good three-year-olds defeat mediocre older horses. A three-year-old champion is harder to intimidate than a young nonentity that not only is afraid to challenge the ancient pecking order but finds comfort in it.

The next time the ordinary three-year-old runs against animals of its own age, it will perform more satisfactorily, all other influences being equal. A smart trainer will not soon enter it in another race against older horses. But if it turns up in such a race, be sure to see how it behaves in the paddock. If it is the least bit uncomfortable, risk no money on its chances. The same advice pertains, by the way, even if a three-year-old has never run against older horses before: If it seems unhappy, take its appearance at face value.

The Winning Stable

After a few visits to paddocks, the racegoer realizes that the best barns send out the best-looking horses. You can make it a rule of thumb: If the horse is bedraggled or fractious and the groom is cranky, you are looking at a pair of losers from a losing outfit. But do not blame the horse. Blame the trainer. And remember the name. When a trainer of that stripe claims a winning horse from a good stable, the horse's winning days are numbered.

As we have declared from time to time, horses read the moods of human beings more accurately than humans read the moods of horses. Discouraged, dispirited stable hands make losers of winners. Eager, happy, interested handlers emit a sense of pleasure and success to which horses react positively. In a setting where the boss pinches pennies at the expense of animals and humans, or is otherwise less than ideally competent, the help is demoralized and the horses go to pieces.

Bear that in mind when evaluating the chances of a horse whose good record was compiled under the guidance of a top trainer. If the animal has now spent more than a week under the roof of a losing trainer, it probably is no longer the same horse.

APPENDIX A:
HOW TO BUY A HORSE

While we were assembling the material for this book, it dawned on us that one of the products of our work should be a whole new category of human awareness. For the first time in the history of the domesticated horse, substantial numbers of persons who lack hands-on experience with the animals will be able to read equine body language. To that extent, they will know more about horses than many experienced horsefolk do.

A warning is in order. Knowledge of horse language is a tremendous asset. But it does not make a horseperson of someone without experience in actual dealings with the animals. To be specific, we hope that no newcomer will assume, having learned the language and no more, that now it is time to purchase and handle a horse.

Let us please agree that nobody who lacks practice in the care, feeding and handling of horses should undertake that kind of activity without expert help. Training in those skills can be obtained wherever horses are domesticated. Readers who have the skills or can employ someone who does, will find this appendix helpful when they go shopping for horses.

The kind of horse you seek will be prescribed, of course, by the uses you have in mind. If, as is most likely, you want a saddle horse and are undecided about the breed, you will enjoy reading the many books and periodicals that celebrate the attractions of each breed. Another consideration surely will be price. The amount of money one can spend dictates the quality of the horse to be bought.

To save time and aggravation, draw up a list of your specifications before you begin shopping. Bear the following truths in mind:

1. A high-strung person should avoid high-strung horses.

2. A relatively inexperienced person is best suited to a docile, experienced horse that needs an absolute minimum of retraining.

3. An easily managed horse is most appropriate for someone who leads a sedentary life.

4. A horse with specialized abilities such as jumping or cutting should be bought only by someone who plans to use those skills. Otherwise, the horse will be wasted and may be unhappy as well at being denied the activity that once was the very center of its existence.

Having decided what kind of animal you want and how much you propose to spend, look for advertisements in publications devoted to the breeds or activities of particular interest. This is almost always more fruitful than a canvass of stables listed in classified telephone directories.

You will quickly locate four or five breeding establishments or training farms that seem to trade in your kind of horse. Call those places. Tell them the specifications. The place you call may have nothing for you but might recommend another that does. After you have five or six real possibilities on your list, stop telephoning and begin traveling. And take a notebook. You will find that it is impossible to remember the distinguishing characteristics of several horses without making careful notes.

As you know very well, the horse's individual history is critically important. This includes the kind of treatment to which it has been subjected during the present phase of its life. Your first clues will come from the physical appearance of the farm or ranch. The neater and cleaner the landscaping, the less evident the junk pile or abandoned tools or peeling paint, the more promising the outlook. The same may be said for the appearance and attitude of the proprietor or manager. The more energetic the salesmanship, the more wary you should become.

For example, it is not at all unusual for a hard-driving salesman to have the horse led out of its stall into the breezeway of the barn and stand it there. But you want to see the animal in the full light of day, not under a shed. If the salesperson is at all reluctant to bring the horse into the sunshine, but seems much more interested in completing the transaction without additional delay, you can confidently climb back into your car and drive off. Horses are not bought so quickly. And horses of approximate soundness and acceptable disposition are put through their complete paces not once but several times—and on different days—before the potential purchaser need make the final decision.

Let us say that the handler brings the horse out of the shed. As it is led into the open, you should follow at a distance of twelve or fifteen feet. You can now tell at once that the animal takes kindly to the halter and lead rope and is not walking lame. But be sure that the handler is using only a loose lead rope hooked to the halter. If a stud chain is used, the horse has a behavior problem. Unless you are prepared to confront such problems, it is time to move to the next farm.

No problems being evident as yet, you have the horse led in a circle. Do its knees and ankles match each other? No bulbous swellings? Hide clean? No bag of bones? Is it following willingly? If it is dancing sideways, is it fractious or merely relieved at being out of the stall? Ask to have it turned loose in a paddock or arena or whatever. A happy horse will now buck, run a while, then toss its head and roll. How does this one roll? Does it sink slowly on one side, roll with some difficulty, struggle to its feet and then roll on the other side? It is not agile enough to roll from one side to the other in an uninterrupted maneuver. Why? Too old? Too arthritic? Too clumsy? Too tired?

Let us assume that the horse has passed all tests so far. Now

comes time to ask the owner some questions. How old is the
horse? Is he registered? What is his particular breeding? How
long have you had him? What have you been doing with him?
Who had him before you? How long was he there? Why was he
sold?

If the horse has changed hands more than three times, beware.
The more so if your questions beget too many vague answers.
When the seller and horse survive this phase of the negotiations,
you should ask if your veterinarian can examine the animal. If
you have no veterinarian who specializes in the care of large
animals, ask the vet who deworms your house cat. But do not use
the one that treats the seller's livestock. It is a shame to cultivate
suspicions of the seller and his horse and his veterinarian, yet
centuries of horse trading have made sharp practice an almost
honorable tradition. A routine deterrent to sharp practice is the
use of one's own veterinarian.

The vet having told you that the horse is pretty much in one
piece and worth what you propose to pay for it, you should now
visit the animal again, this time to groom it and handle it in a
walking ring. This will give you an idea as to the ease with which
you will be able to form a good working relationship. After that,
use a saddle and bridle and revisit the ring for a walk, trot and
slow canter. Everything still satisfactory? Tell the owner that you
kind of like the horse and want to think about it.

If the owner now seems awfully impatient to close the deal and
begins pushing, you are entitled to turn slightly paranoid. Why is
he in such a hurry?

Let us repeat that horse trading is not a pastime for the inno-
cent. Compare the purchase of an elderly horse to that of an
elderly used car, but complicated by the fact that you probably
know more about the intricacies of cars and the habits of car
salesmen than about horses and horse traders. Why is this seller
in a rush when all the traditions of his field make the purchase
of a horse as protracted and ceremonious as that of a gem in an
Arabian bazaar? Does this animal have some behavioral or physical
flaw that shows up only in special circumstances and has eluded
not only your attention but your veterinarian's?

Take your leave, promising to return soon. But on the next visit make your appearance at a new time of day. Have you been coming in the afternoon? Try the morning. See if something is going on in the establishment that had not been evident before. For example: Do you suddenly see five or ten pathetic-looking horses that seem en route to the rendering plant but were out of sight during previous visits? What's going on here?

No such unsettling surprises having greeted you, take the horse out again for a real test. Turn it loose in the paddock for a while. Then get it for saddling. Ask permission to take it for a half-hour ride off the premises. See if the horse is tractable and willing when on the road. Work him enough to warm him. Does he begin limping or breathing noisily?

The final test, especially if you plan to join a riding club for cross-country trips, is the trailer. Does this horse go into a trailer without difficulty? If not, you have a deadly serious problem to solve before you can join the cross-country club. The older the horse and the more energetic its objection to the trailer, the deeper the problem and the less probable its solution.

The horse passes the test. You want to buy. The seller supplies the animal's original registration papers (if the horse is represented as possessing such eminence). On the back of the papers should be the dated signature of everybody who has ever owned the animal.

Lead the horse away and live happily ever after!

APPENDIX B:
HOW TO BUY A RACE HORSE

This brief and highly specialized afterword emphasizes some points made previously. We repeat the ideas here because we want one more opportunity to attract the attention of the rare human being able to implement them.

We shall not dwell on the traditions that govern the advertising, promotion and auction sale of race horses. But we urge the reader to expect a shell game. Nobody should spend a penny on a race horse without the advice of a trusted expert.

Assuming that you have found such an advisor and that your trust is well founded, we congratulate you. We now ask you to consider—or reconsider—something we said earlier in this book.

The Thoroughbred, Standardbred or Quarter-horse foal most likely to become a good racer is one that not only is suitably bred and sound of body and balance but occupies the privileged position of Number One foal. That is, its mother is the lead mare. Long before it is weaned, and even longer before it goes to auction, the lead foal becomes accustomed to deference from others of its generation. You and your advisor should go to breeding farms and observe the mares and foals in their pastures. See which ones run the show. Find out the name and background of the lead mare and of the foal's sire. Return at intervals to see whether the foal remains sound and takes comfortably to its privileges. Watch it, or have it watched, during and after weaning. Does it retain its leadership now that Mom is no longer around?

That is the kind of foal to buy. If your reasoning and purposes are not too conspicuous and, therefore, do not drive up the price, you may be able to buy the youngster as a weanling, rather than wait another year and pay an inflated bill at auction.

Repeat: The race horse to buy is the Number One foal, whose subsequent experiences have done nothing to disabuse him of his sense of power. Someone interested in buying more than one youngster at a time should buy nothing but Number One foals. It can be done.

We wish you luck.

INDEX